Studies in Logic
Logic and Bounded Rationality
Volume 102

Depth-bounded Reasoning
Volume 1
Classical Propositional Logic

Volume 95
Transparent Intensional Logic. Selected Recent Essays
Marie Duží, Daniela Glaviničovà, Bjørn Jespersen and Miloš Kosterec, eds

Volume 96
BCK Algebras versus m-BCK Algebras. Foundations
Afrodita Iorgulescu

Volume 97
The Logic of Knowledge Bases, Second Edition
Hector Levesque and Gerhard Lakemeyer

Volume 98
Classification Theory. Second Edition
Saharon Shelah

Volume 99
Theories of Paradox in the Middle Ages
Stephen Read and Barbara Bartocci, eds

Volume 100
The Fallacy of Composition: Critical Reviews, Conceptual Analyses, and Case Studies
Maurice A. Finocchiaro

Volume 101
The Logic of Partitions. With Two Major Applications
David Ellerman

Volume 102
Depth-bounded Reasoning. Volume 1: Classical Propositional Logic
Marcello D'Agostino, Dov Gabbay, Costanza Larese and Sanjay Modgil

Studies in Logic/Logic and Bounded Rationality Series Editors
Marcello D'Agostino, Dov Gabbay, Costanza Larese and Sanjay Modgil

Depth-bounded Reasoning
Volume 1
Classical Propositional Logic

Marcello D'Agostino
Dov Gabbay
Costanza Larese
Sanjay Modgil

© Individual authors and College Publications, 2024
All rights reserved.

ISBN 978-1-84890-442-2

College Publications
Scientific Director: Dov Gabbay
Managing Director: Jane Spurr

http://www.collegepublications.co.uk

All rights reserved. No part of this publication may be reproduced, stored in a retrieval system or transmitted in any form, or by any means, electronic, mechanical, photocopying, recording or otherwise without prior permission, in writing, from the publisher.

Contents

Preface		ix
Introduction		xi
Prequel		xvii

I Depth-bounded Boolean logics   1

1 The 0-depth layer   3

- 1.1 An informational view of classical logic 3
  - 1.1.1 The cost of thinking 3
  - 1.1.2 Approximation systems 7
  - 1.1.3 Actual vs virtual information 9
  - 1.1.4 The depth-bounded project 12
- 1.2 Informational semantics 14
  - 1.2.1 The need for an informational semantics 14
  - 1.2.2 Informational vs 3-valued semantics 18
  - 1.2.3 Informational vs intuitionistic semantics 22
- 1.3 Constraint-based semantics 24
  - 1.3.1 The informational meaning of the logical operators . 24
  - 1.3.2 Meaning via negative constraints 25
  - 1.3.3 The single candidate principle 29
- 1.4 Intelim sequences 35
  - 1.4.1 The emergence of inference rules 35
  - 1.4.2 Intelim sequences 37

		1.4.3 Another version of the single candidate principle	44
		1.4.4 Soundness and completeness	46
		1.4.5 Subformula property and tractability	47
	1.5	Non-deterministic semantics	59
		1.5.1 Valuation systems	59
		1.5.2 Lack of a finite valuation system for the 0-depth logic	60
		1.5.3 Non-deterministic tables	61
		1.5.4 Expressive completeness	63

2 Depth-bounded deduction — 65

- 2.1 Weak depth-bounded approximations 65
  - 2.1.1 Virtual information 65
  - 2.1.2 Virtual space and search space 69
  - 2.1.3 Weak k-depth approximations 73
  - 2.1.4 C-intelim tableaux 75
  - 2.1.5 Normal and quasi-normal tableaux 80
- 2.2 Strong depth-bounded approximations 87
  - 2.2.1 Depth-bounded information states 87
  - 2.2.2 C-Intelim hyper-sequences 88
- 2.3 Non-contamination 98
  - 2.3.1 The contamination problem. 99
  - 2.3.2 Variable sharing and non-contamination 103

II Applications of depth-bounded reasoning — 105

3 Rational non-monotonic reasoning — 107

- 3.1 Introduction 107
- 3.2 Argumentation and non-monotonic logic 110
  - 3.2.1 Argumentative characterisations of NML 111
  - 3.2.2 From argumentation to dialogical characterisations of non-monotonic reasoning 117
- 3.3 Dialectical classical logic argumentation 119
  - 3.3.1 K-depth classical logic arguments 119
  - 3.3.2 Non-monotonic reasoning, argumentation and rationality 121

	3.3.3 Dialectical acceptability and k-depth argumentation frameworks 125

- 3.4 Conclusions and further remarks 134

4 Problems in the philosophy of logic   **137**
- 4.1 Is logic analytic? 137
  - 4.1.1 Kant 138
  - 4.1.2 Frege 140
  - 4.1.3 Logical empiricism 142
  - 4.1.4 Carnap and Quine 143
- 4.2 Is logic tautological? 146
  - 4.2.1 The scandal of deduction 147
  - 4.2.2 Semantic information 151
  - 4.2.3 The BHC paradox 153
- 4.3 Attempts to explain the scandals 155
  - 4.3.1 Logical empiricism 155
  - 4.3.2 Wittgenstein 158
- 4.4 Attempts to reject the scandal 161
  - 4.4.1 Frege 161
  - 4.4.2 Hintikka 165
- 4.5 The depth-bounded approach 171
  - 4.5.1 Defusing the scandal 171
  - 4.5.2 Semantic information revisited 176

Conclusion   **181**

A Proofs and algorithms   **189**
- A.1 Proof of Proposition 1.4.3 189
- A.2 Proof of Proposition 1.4.15 190
- A.3 Tractability 191
- A.4 Proof of Proposition 2.1.20 195
- A.5 Proof of Lemma 2.1.17 196
- A.6 The full expand algorithm 197

Bibliography   **199**

Index   **219**

Preface

The "cost of reasoning", i.e. the cognitive and computational effort required by non-ideal, resource-bounded agents (whether human or artificial) in order to process available information, is a crucial issue in philosophy, AI, economics and cognitive science. Accounting for this fundamental variable in modelling inference, argumentation and decision-making for limited real-world agents is one of the most important and difficult challenges in the theory of rationality. It is also pivotal to significant advances in the field of "hybrid intelligence", i.e. the combination of human and machine intelligence, one of the most promising directions of research in contemporary human-centered AI. With this book, we intend to launch a series that, under the general title of "Logic and Bounded Rationality", aims to create a community of researchers from several areas that wish to cooperate towards a systematic *logical* view of rational argumentation and deliberation under limited resources.

With this book we start a project that aims to put forward a specific "informational" view of bounded reasoning that stems from a long standing research program. In this first volume we shall focus on reasoning in classical propositional logic. In future volumes we plan to cover first-order logic as well as a variety of non classical reasoning systems. Proposals for other projects related to the general theme of "logic and bounded rationality" will be very welcome.

We wish to thank all members of the L.U.C.I. (Logic, Uncertainty, Information and Computation) Lab based in the University of Milan, and especially Hykel Hosni and Giuseppe Primiero, for many interesting discussions and fruitful exchange of ideas on the general project as well as on the contents of this book. Special thanks are due to Luciano Floridi and Marcelo

Finger, for their work on the philosophical and computational ideas underlying the depth-bounded paradigm, as well as to Alejandro Solares-Rojas, for making important contributions to extending it to non-classical logics, and to Jane Spurr for her valuable and enduring support. The first author is deeply indebted to his friend and mentor Marco Mondadori (1945-1999) from whom all this started.

This research was funded by the Department of Philosophy "Piero Martinetti" of the University of Milan under the Project "Departments of Excellence 2023–2027" awarded by the Ministry of University and Research (MUR). We also thankfully acknowledge the support of the Italian Ministry of University and Research (PRIN 2022 project n. 2022ZLLR3T).

M. D'A., D.M. G., C. L., S. M.

Introduction

Many artificial intelligence and applied computer science systems contain a key deductive component. However, establishing whether a certain conclusion follows from given premises is often beyond human cognitive resources and, in general, is also computationally hard. There is a need for a theoretical discipline of feasible deduction that can serve as a practical component in all of these applications. The main problem, in a nutshell, is that standard logical systems model *logically omniscient* agents — able to correctly recognize *all* consequences of their assumptions according to the system in use — but provide no sensible means to account for the cost of inferring them. In contrast, a single practical agent (whether human or artificial) using a logic L cannot be expected to effectively perform all the correct inferences of L, but only those that are within the reach of its limited resources. In practice, a suitable approximation method is needed to model realistic agents that reason according to their computational limitations.

We do not think that just limiting the number of proof steps that an agent can perform to a certain finite number is an answer to this problem. We need to decompose deduction systems into *logically meaningful layers* in the sense that they respect criteria of rationality within each given "layer", forming a hierarchy in which the lower layers *naturally* require fewer computational steps. The same applies to any solution that takes the amount of resources consumed in solving a problem as a parameter to measure its difficulty.[1] The hierarchy of logics mentioned in Gabbay & Woods (2001) was

---

[1] A similar problem was considered by Gabbay & Owens (1991) in connection with real-time temporal logics. We cannot define a real-time system as a system able to handle an application in real time. This makes the notion of "real time" dependent on the power of computers. What is considered today as "not real-time" might be considered as "real-time"

also determined by the resources available to the agents,[2] and so it was also a resource-defined hierarchy and not one defined by logically meaningful layers. This leads us to the following:

> **Approximation problem**: Given a logic L, can we define *in a natural way* a hierarchy of approximating logical systems that converge to L in such a way that these approximations satisfy minimal rational requirements and can be sensibly used as formal models of the deductive power of resource-bounded agents?

Robust solutions to this problem are likely to have a significant practical impact in many research areas — from economics, to philosophy, artificial intelligence and cognitive science — wherever there is an urgent need for more realistic models of deduction.

Despite its significance, both practical and theoretical, the approximation problem has been surprisingly neglected in the logical literature.[3] A first reason is the difficulty of finding solutions which are independent of the choice of a specific formalism. A deeper reason is the received view that logic is *informationally trivial*: the information carried by the conclusion is (in some sense) "contained" in the information carried by the premises. This view completely ignores the computational effort required to obtain the conclusion. In fact, there is a body of results showing that most interesting logics are either undecidable or (very likely to be) intractable.

A typical response to this objection is that logical systems are *idealizations* and, therefore, they are not intended to model the actual inferential

---

tomorrow. What is needed is a logical characterization of real time.

[2]"At the bottom of this hierarchy are individual human beings with minimal efficient access to institutionalized databases. Next up are individual human beings who operate in institutional environments — in universities or government departments for example — which themselves are kinds of agents. Then too, there are teams of such people" (Gabbay & Woods, 2001, p. 145).

[3]It has indeed received some attention in Computer Science and Artificial Intelligence (Cadoli & Schaerf, 1992; Dalal, 1996, 1998; Crawford & Etherington, 1998; Massacci, 1998; Sheeran & Stålmarck, 2000; Finger, 2004a,b; Finger & Wassermann, 2004, 2006; Finger & Gabbay, 2006; Lakemeyer & Levesque, 2020), but comparatively little attention has been devoted to embedding such efforts in a systematic proof-theoretical and semantic framework.

power of real agents. Thus, logic is informationally trivial for ideal agents, but not for the real, resource-bounded agents that operate in practice.

Granting that a certain amount of idealized assumptions is necessarily involved in any scientific model, we look for an approach to logical systems that allows for increasing *degrees of idealization*. As Gabbay and Woods put it:

> A logic is an idealization of certain sorts of real-life phenomena. By their very nature, idealizations misdescribe the behaviour of actual agents. This is to be tolerated when two conditions are met. One is that the actual behaviour of actual agents can defensibly be made out to approximate to the behaviour of the ideal agents of the logician's idealization. The other is the idealization's facilitation of the logician's discovery and demonstration of deep laws (Gabbay & Woods, 2001, p. 158).

This should not be intended as a plea for a more descriptive approach to the actual inferential behaviour of agents, which takes into account their cognitive biases or distortions. Even from a prescriptive viewpoint, the requirements that logic imposes on agents are too strong and there is a need for less demanding approximating models. In this view, idealized logics are not dismissed as descriptively inadequate, but still play a crucial role as limiting normative theories to which approximating models should converge. Our approach can be seen as an evolution of the algorithmic view of logical systems put forward in Gabbay (1991), according to which:

> A logical system L is not just the traditional consequence relation $\vdash$ (monotonic or non-monotonic), but a pair $(\vdash, S_\vdash)$, where $\vdash$ is a mathematically defined consequence relation (i.e. the set of pairs (Δ, Γ) such that $\Delta \vdash \Gamma$) [...] and $S_\vdash$ is an algorithmic proof system for generating all those pairs. Thus, according to this definition, classical propositional logic $\vdash$ perceived as a set of tautologies together with a Gentzen system $S_\vdash$ is not the same as classical logic together with the two valued truth-table decision procedure $T_\vdash$ for it. In our conceptual framework, $(\vdash, S_\vdash)$ is *not the same logic* as $(\vdash, T_\vdash)$ (Gabbay, 1991, p. 185).

Now we require that a logical system L consist not only of a consequence relation with an algorithmic proof system for it, but also a *logically meaningful* definition of how L can be approximated in practice by realistic (not logically omniscient) agents.

This sophisticated way of dealing with the discrepancy between theoretical assumptions and practical applications cannot be easily accommodated — and in some cases is not even conceivable — in the framework of traditional mathematical logic. Neither does it belong to the repertoire of the non-classical logics that have been proliferating in artificial intelligence and computer science. The interesting contributions that have been made are somewhat scattered in the literature, and have not yet given rise to a shared paradigm in the pure and applied logic community. It is the ambition of this book to set up such a paradigm by proposing a systematic approach to the approximation problem that should be appealing to logicians, philosophers, computer scientists, as well as to economists and cognitive scientists working on realistic models of rationality.

In this volume we shall address the problem by outlining a novel "informational view" of propositional classical logic which is based on ideas and results that have been anticipated in a series of papers (D'Agostino, 2010, 2013a,b, 2014a,b, 2015, 2016, 2019; D'Agostino & Floridi, 2009, 2015; D'Agostino et al., 2013, 2020), but are combined and extended here for the first time into an overall systematic picture. The book has both philosophical and computational content, intertwined in a way that we hope will convince philosophers that their conceptual analyses may be considerably refined taking into account notions and methods from more mundane subjects such as engineering and computer science. Conversely, we hope to convince applied scientists that philosophy may well provide important insights for the problems they typically address.

The conceptual core of our novel view of classical logic consists of a kind of "informational semantics" for the classical operators whose proof-theoretical presentation is a system of classical natural deduction (Mondadori, 1989a; D'Agostino, 2005; D'Agostino et al., 2020) that, unlike the standard Gentzen-style systems, provides a simple means for measuring the "depth" of inference. We argue that this approach may be apt to provide an adequate solution to the approximation problem in that it leads to defining, in a simple and intuitive way, a hierarchy of tractable depth-bounded deduction systems. As recent applications to non-monotonic reasoning suggest (D'Agostino & Modgil, 2018a), this hierarchy appears to be a plausible model for representing rational agents with increasing, albeit limited, cognitive and computational resources.

This book is organised into a brief prequel followed by two main parts and a short concluding chapter. The Prequel gives an intuitive introduction to the key ideas by means of familiar examples based on the popular *sudoku* puzzle. Part I provides a detailed presentation of the depth-bounded paradigm in the case of classical propositional logic. Part II discusses its applications to non-monotonic reasoning (via formal argumentation theory) and to some enduring problems in the philosophy of logic. The concluding chapter provides a discussion of ideas and methods that are most closely connected to the semantic and proof-theoretical framework presented here, as well as the state of the art of the research programme that stems from it.

Prequel: the logic of sudoku

> If you want to go down deep,
> you don't need to travel far;
> indeed you don't need to leave
> your most immediate and
> familiar surroundings. [...]
> [...] How small a thought it takes
> to fill a whole life
>
> Ludwig Wittgenstein

In this prequel we provide an intuitive introduction to the central idea of this book, namely that a measure of the difficulty of logical deduction can be obtained by focusing on the essential use of "virtual information" and on the depth at which this use is required to infer a given conclusion. We shall illustrate this idea by means of familiar examples based on the popular *sudoku* puzzles.[4] In its most general version, a sudoku puzzle requires inserting positive integers between 1 and n^2 in a grid of $n^2 \times n^2$ elements subdivided into n^2 quadrants ("houses"), each containing n^2 cells, in such a way as to satisfy the following constraint: every number must be contained *exactly once* in each row, column and house. The usual sudoku puzzles found in the newspapers are 9×9 (i.e., $n = 3$). In its general for-

---

[4]The connection between easy sudoku puzzles and easy inferences in classical propositional logic (those which do not require the use of virtual information) was pointed out in D'Agostino & Floridi (2009, p. 299). Here, our purpose is only to illustrate this connection and provide a familiar context in wich this distinction can be easily grasped. By now, there is a rich and interesting literature on the logical analysis of Sudoku and we do not even attempt to provide adequate references.

mat, solving a sudoku puzzle is an NP-complete problem, hence equivalent to the satisfiability problem for propositional logic, and most likely to be intractable (Yato & Seta, 2003).

Consider the top grid in Figure 1(a) and focus on the cell $r2c7$ (row 2, column 7). The sudoku rules *force* us to insert 2 in this cell as an *immediate result* of the following constraints: (i) the cell cannot contain 3, 5, 6, 7, 8, because these numbers are already contained in row 2; (ii) it cannot contain 4 because this is already contained in column 7; (iii) it cannot contain 1, 6, 9 because these numbers are already contained in the same house.

From these constraints, it follows that there is *only* one candidate for cell $r2c7$, which is 2 (candidates for a cell are shown in small font). Thus we can immediately insert it in the grid as shown in the second grid of Figure 1(a). This is a typical "easy" sudoku step that can be expressed in propositional logic as follows:

$$\frac{\begin{array}{c} C_{27}(1) \vee \cdots \vee C_{27}(9) \\ \neg C_{27}(1) \\ \neg C_{27}(3) \\ \vdots \\ \neg C_{27}(9) \end{array}}{C_{27}(2)}$$

where "$C_{ij}(n)$" means that $ricj$ contains n, "$\neg$" is the usual symbol for negation and "$\vee$" the usual symbol for disjunction. This is a typical reasoning pattern that "dictates" a certain alternative by excluding one by one all the others, and is sometimes called "the single candidate principle". The special case illustrated in Figure 1(a) is called "naked singles". We can continue filling up the grid by means of such simple steps. For example, there is again a single candidate (9) for cell $r2c5$ in the second grid of Figure 1(a), and so its content is dictated with no need for further analysis, which leads us to the third grid, and so on.

Another special case of the single candidate principle is called "hidden singles" and is illustrated in Figure 1(b). In the top grid, if we look only at the constraints concerning cell $r1c2$, there seem to be five candidates for it. However, all of them except 9 are not real candidates. For, we know that 9 must occur in the first house, but it cannot occur in rows 2 and 3, and it

xix

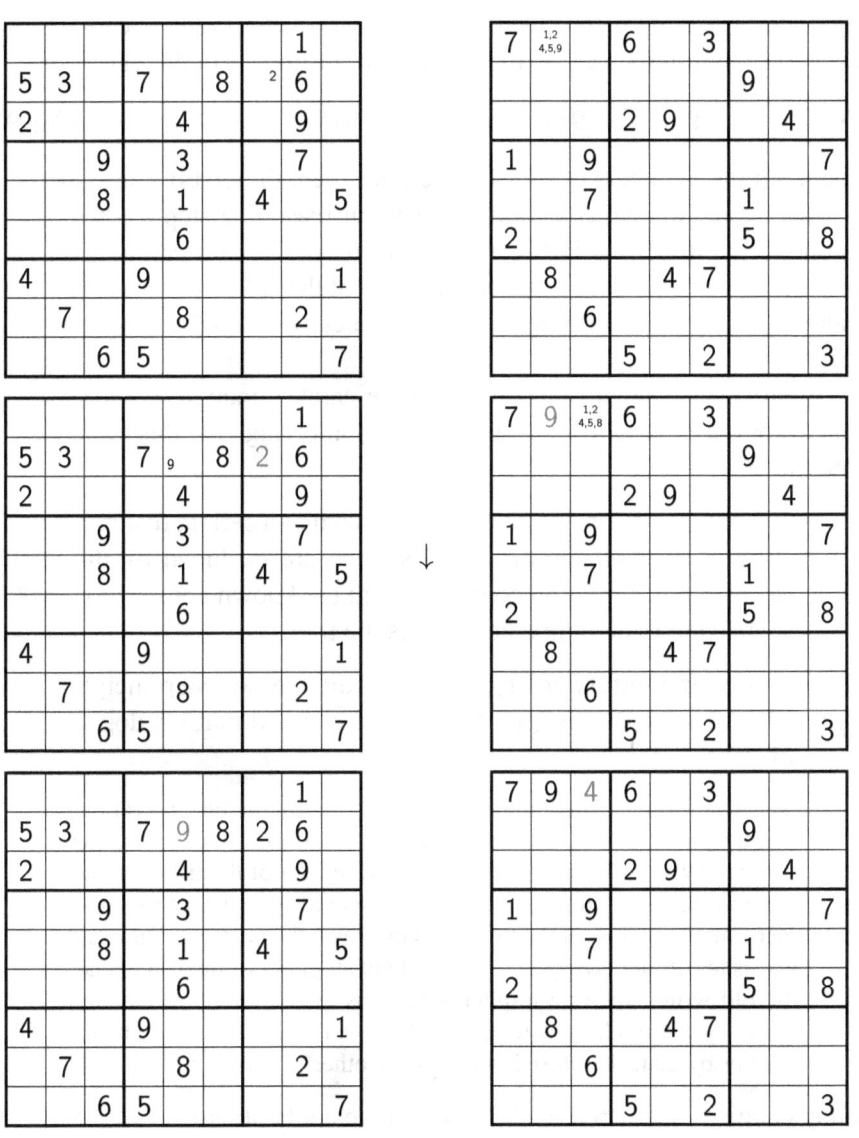

(a) Naked singles (b) Hidden singles

Figure 1: Easy sudoku steps.

cannot occur in column 3 either. From the point of view of propositional logic, the reasoning pattern is the same as the previous one, except that the initial disjunction focuses on the fact that 9 must occur in the first house:

$$C_{12}(9) \vee C_{13}(9) \vee C_{21}(9) \vee C_{22}(9) \vee C_{23}(9) \vee C_{31}(9) \vee C_{32}(9) \vee C_{33}(9).$$

Again, all the members of this disjunction except the first one are ruled out by some of the known constraints. A similar reasoning step leads from the second to the third grid in Figure 1(b). Here, the key observation is that 4 must occur in row 1, but cannot occur in any of the cells $r1c7$–$r1c9$ (because 4 already occurs in the house) and cannot occur in $r1c5$ (because it already occurs in the column).

In an "easy" sudoku puzzle, if we are careful enough, we are usually able to complete the grid using an ink pen by applying only this basic logical pattern:

> *Single Candidate Principle.* We know that a certain disjunction is true; all the disjuncts but one are excluded by the available information combined with the known constraints; hence the remaining disjunct must be true.

This is a very straightforward and time-honoured reasoning principle. Chrysippus (III Century BC) is reported to have claimed that his dog was well aware of it:

> And according to Chrysippus, who shows special interest in irrational animals, the dog even shares in the far-famed "Dialectic". This person, at any rate, declares that the dog makes use of the fifth complex indemonstrable syllogism when, arriving at a spot where three ways meet, after smelling at the two roads by which the quarry did not pass, he rushes off at once by the third without stopping to smell. For, says the old writer, the dog implicitly reasons thus: "The creature went either by this road, or by that, or by the other: but it did not go by this road or by that: therefore it went by the other" (Sextus, 1933, p. 41).

Many centuries later, Sherlock Holmes repeatedly stresses the importance of this principle when he lectures his unsophisticated assistant Watson on his "elementary" trains of thought:

> How often have I said to you that when you have eliminated the impossible, whatever remains, however improbable, must be the truth?

[...] Eliminate all other factors, and the one which remains must be the truth (Conan Doyle, 1981, p. 111).

However, Holmes never instructs Watson that the single candidate principle is by no means sufficient to solve all problems that may arise in propositional logic. This is already apparent when we move from the easiest sudoku puzzles to the more difficult ones. Consider, for example, the harder one in Figure 2. If we focus on row 4 of the top grid, we observe that 1 is a candidate in cells $r4c4$, $r4c6$, $r4c7$. At this stage, there is no way of applying the single candidate principle and we need to keep track of alternate *hypotheses* and analyse their consequences. *Suppose* first that 1 occurs in $r4c4$, then the single candidate principle allows us to continue filling the grid by the following steps: $r8c4$ must contain 6, $r8c8$ must contain 1, $r6c8$ must contain 6 (it is the only cell in the column with 6 as candidate), $r1c8$ must contain 3 (for the same reason as in the previous step), $r1c7$ must contain 4. Now, *suppose* instead that 1 does not occur in $r4c4$, and consider the next alternative, that it occurs in $r4c6$. By following a similar chain of "single candidate" steps, we find that they also lead to the same conclusion that $r1c7$ must contain 4. Finally, suppose that 1 does not occur in $r4c6$ either. Then it must occur in $r4c7$ and, by a similar chain of easy steps we obtain once again the same conclusion that 4 must occur in $r1c7$. Hence, no matter what the location of 1 in row 4 actually is, we have inferred that the content of $r1c7$ must be 4. This is an example of what is usually called "case reasoning" and, can be called the *consensus principle*:

> *Consensus principle.* We consider a set of mutually exclusive and collectively exhaustive alternatives; if all those that do not lead to impossible consequences agree on a certain conclusion, than this conclusion holds independently of which of the alternatives is the true one.

A special case of the consensus principle arises when there are only two complementary alternatives. Figure 3 shows that a combination of this special case with the Single Candidate Principle is sufficient to simulate the general consensus principle (where the dotted lines stand for arguments that lead all to the same conclusion except for those that lead to a dead end). This logical pattern is essentially different from the one used in the previous

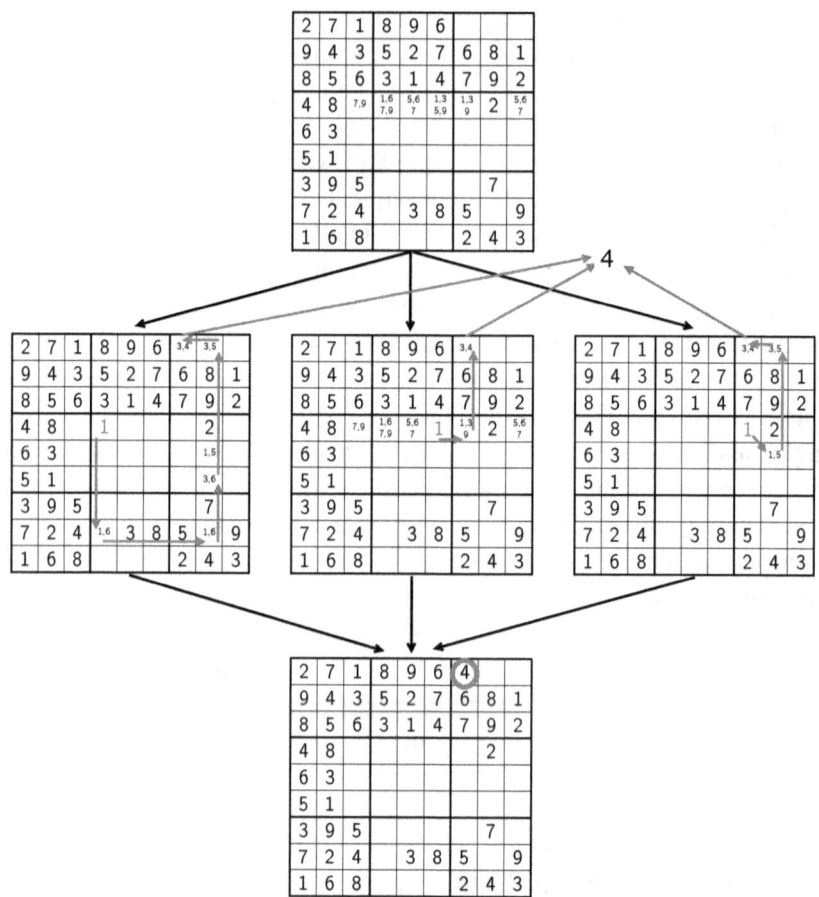

Figure 2: Harder Sudoku steps (using virtual information).

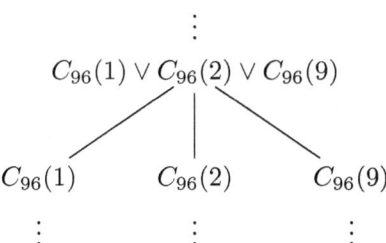

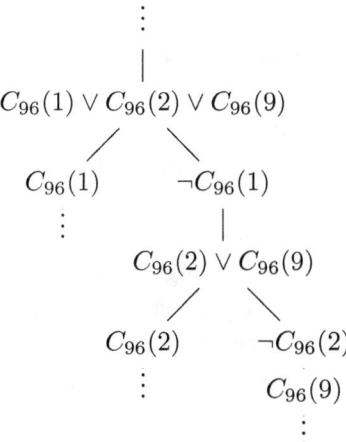

Figure 3: Simulating n-ary branching by binary branching.

examples. In the "single candidate" principle we process only *actual information*, i.e. information that we actually possess, with no need for making alternative hypotheses to simulate the missing information about a cell. By contrast, in the case reasoning used for the example in Figure 2 we essentially need to process also *virtual information*, i.e., all admissible (given the constraints) extensions of the actual information state concerning the column containing 1 in $r4$, and recognize that in all these extensions the content of $r1c7$ is 4. When the consensus principle is involved, we need to process information that we do not actually possess and is not implicitly contained in the information that we possess. Therefore, we are not usually able to complete the sudoku puzzle without using pencil and rubber to annotate the candidates, keep track of the alternative suppositions together with their consequences, and remove such virtual information when a consensus is reached.

It may be the case that, before the alternative hypotheses agree on the content of a cell, further case analysis is required involving further virtual information. The depth at which such nested use and processing of virtual information is required provides a simple and plausible measure of the difficulty of solving a sudoku puzzle. The general structure of such complex reasoning patterns is illustrated in Figure 4, where the dotted lines represent sequences of applications of the single candidate principle, the downward branchings represent the introduction of "virtual information" concerning the content of a cell and the upward branchings applications of the consensus principle.

In each new branch we simulate an information state that is essentially richer than the actual one, introducing virtual information that is not even implicitly contained in it. In the path ending with $\times$ the process leads to a "dead end" (an impossible information state that violates the sudoku rules). In the others the alternative virtual states reach a consensus on the content of some specific cell. These more complex patterns require an *objective* effort to be performed by any realistic agent; effort that grows with the depth at which the nested use of virtual information is required.

We maintain that the difference between reasoning patterns that are based on the information that we actually possess, such as the single candidate principle, and those that essentially require the introduction of virtual information is non-trivial from the philosophical, computational and cogni-

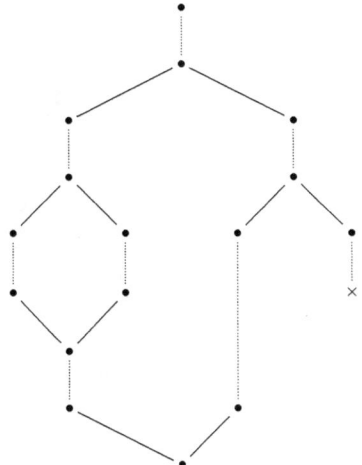

Figure 4: Nested virtual information.

tive viewpoints. In this book we show that clarifying this distinction, and all its implications, leads to interesting measures of difficulty of logical reasoning that may help to provide more realistic models of rational agents.

Part I

Depth-bounded Boolean logics

Chapter 1

The 0-depth layer

1.1 An informational view of classical logic

1.1.1 The cost of thinking

According to the received view (see also Chapter 4) logic is *informationally trivial*. Deductive inference is commonly described as being "tautological", in the sense of being *uninformative* or *non-ampliative*: the information carried by the conclusion is (implicitly) contained in the information carried by the premises. We know, however, that extracting such implicit information from a given set of premises, e.g., a knowledge base or a set of axioms of a mathematical theory, in most interesting cases can be *very* hard. The fundamental question is "in what sense the conclusion is implicitly contained in the premises"?

(1.1) Do reasoning agents possess the information that the conclusion of a valid inference, in a given logic L, is true *whenever* they possess the information that its premises are true?

By "reasoning agent", in this book, we mean any system — no matter whether human, artificial or a combination of the two — provided with inferential capabilities.

The answer to (1.1) depends on what we mean by "possessing" a piece of information. If we intend that this information is, at least in principle, *available* to the agent, the answer crucially depends on the mastery of a decision procedure for the logic L. Then, the lack of such a general decision procedure implies that this answer is a loud "NO" for classical first-order logic: there is no guarantee that we can *always* get ourselves into a position in which we can recognize the truth of a valid consequence A of a set Γ of sentences whenever we recognize the truth of the sentences in Γ. A positive answer to (1.1) would imply that we may "possess" a piece of information, and yet be unable, even in principle, to recognize that we possess it. Thus, there might be in some cases no observable difference between an agent a who implicitly "possesses" a certain piece of information, corresponding to a remote logical consequence of information that she possesses explicitly, and an agent b who does not because her own information is insufficient to establish this consequence. So, the kind of implicit information that may justify the claim that logic is informationally trivial is not fully manifestable, which is quite at odds with the commonsense notion of information and explains the intuitive negative answer to the fundamental question (1.1).[1] There is no claim that *all* kinds of information should be manifestable in this sense, but only that a notion of manifestable information is eminently worthy of logical investigation.

By contrast, the availability of a decision procedure at first sight appears to support the claim that classical propositional logic is indeed tautological. But even in this restricted domain the intuitive answer to (1.1) is in the negative. The theory of computational complexity[2] tells us that the decision problem for Boolean logic is coNP-complete (Cook, 1971), that is, among the most difficult problems in coNP. Even if yet unproven, the conjecture that there exists no feasible decision procedure for such problems is widely believed to be true. This means that we cannot expect a non-ideal rational agent to *always* be capable of recognizing *in practice* that a certain conclusion follows from a given set of premises. Even if the implicit information contained in the premises is available in principle, it is not always available in practice and so it is not fully manifestable.

---

[1]This is analogous to Dummett's manifestability requirement for the implicit knowledge of meaning: "Implicit knowledge cannot, however, meaningfully be ascribed to someone unless it is possible to say in what the manifestation of that knowledge consists: there must be an observable difference between the behaviour or capacities of someone who is said to have that knowledge and someone who is said to lack it" (Dummett, 1978, pp. 217).

[2]See (Stockmeyer, 1987) for an introduction.

1.1. AN INFORMATIONAL VIEW OF CLASSICAL LOGIC

Things do not get better if we consider the best known subclassical propositional logics, such as intuitionistic, relevance and linear logic. As for intuitionistic logic, Richard Statman (1979) proved that it is PSPACE-complete (as is its pure implication fragment),[3] and so — given that NP $\subseteq$ PSPACE and that the inclusion is believed to be strict — its computational complexity is likely to be worse than that of classical logic. As for Relevance Logics, Alasdair Urquhart proved that the main systems E (Entailment), R (Relevance Logic) and T (Ticket Entailment) (Anderson & Belnap Jr, 1975) are all undecidable (Urquhart, 1984) and that the computational complexity of the main decidable subsystems is not better than that of Boolean logic (Urquhart, 1990). For example, the implication fragment of R is ESPACE complete and the fragment known as *first-degree entailment* — characterized by Belnap's 4-valued logic (Belnap Jr., 1976, 1977) — is coNP-complete. Finally, the full system of Linear Logic is undecidable (Lincoln, 1995), its multiplicative-additive fragment is PSPACE complete (Lincoln *et al.*, 1992), its multiplicative fragment is NP-complete (Kanovich, 1992), and even the constant-only fragment is NP-complete (Lincoln & Winkler, 1994). Finally, shifting to non-monotonic reasoning, it was recognized early on that "a general property of non-monotonic inference is that its computational complexity is higher than the complexity of the underlying monotonic logic" (Cadoli & Schaerf, 1993, p. 128).[4]

Thus, there is a tension between the traditional view that logic is informationally trivial and the results showing that most interesting logics are either undecidable or (very likely to be) intractable. How can logic be, at the same time, informationally trivial and computationally hard? This issue was recognized early on by philosophers and logicians (see Chapter 4) as conceptually embarrassing, and carried over to epistemic and doxastic logic, i.e. the attempts to extend propositional logic by augmenting its language with suitable operators for expressing knowledge and belief. In this context, it

---

[3] More precisely (Statman, 1979) shows that intuitionistic propositional logic can be reduced to its implication fragment and that the latter is PSPACE-hard. It then follows from (Ladner, 1977), where it is shown that S4 in in PSPACE and from the well-known polynomial translation of intuitionistic propositional logic into S4, that the decision problem for intuitionistic propositional logic is PSPACE-complete.

[4] Assuming "supraclassical" non-monotonic logics (e.g., Reiter's Default Logic (Reiter, 1980)) that augment the classical inferences from a set Γ with default or 'defeasible' inferences that may subsequently be withdrawn upon expanding Γ.

is widely known as the "problem of logical omniscience" (Levesque, 1984; Fagin *et al.*, 1995b; D'Agostino, 2010; Artemov & Kuznets, 2014; Hawke *et al.*, 2020a, see also Chapter 4 below) and has been the object of intensive research that, despite significant progress, does not yet appear to have reached a stable solution.

The tension is widely acknowledged also in cognitive science. In his book *Minimal Rationality*, Christopher Cherniak, as early as in 1986, maintained that a theory of feasible and appropriate logical inferences constitutes an important component of the theory of rationality, but an *unconstrained* logic is by itself irrelevant to the psychology and epistemology of reasoners. The main problem is that standard logical systems model logically omniscient reasoners and provide no means to account for the "cost of thinking":

> Until recently, philosophy has uncritically accepted highly idealized conceptions of rationality. But cognition, computation and information have costs; they do not just subsist in some immaterial effluvium (Cherniak, 1986a, p. 3).

In the same vein, Hector Levesque's plea for a less idealized view of logic is worth quoting in full:

> What does logic have to do with computational approaches to the study of cognition? Very little, it would seem. For one thing, people (even trained logicians) are unquestionably very bad at it, compared to their skill at (say) reading or recognizing tunes. And computers have a hard time with it too: the computational activity that goes with logic, theorem proving of some sort, appears to be computationally intractable. Given its apparent difficulty, it seems quite unlikely that logic could be at the root of normal, everyday thinking. What I want to suggest here, however, is that rather than closing the book on logic (except perhaps as a mathematical tool for theoreticians so inclined), these facts force us into a less idealized view of logic, one that takes very seriously the idea that certain computational tasks are relatively easy, and others more difficult (Levesque, 1988, p. 355).

This is essentially the same problem raised, in the context of probability

1.1. AN INFORMATIONAL VIEW OF CLASSICAL LOGIC

theory, by Leonard Savage in his *Difficulties in the theory of personal probability*:

> A person required to risk money on a remote digit of π in order to comply fully with the theory [of personal probability] would have to compute that digit, though this would really be wasteful if the cost of computation were more than the prize involved. For the postulates of the theory imply that you should behave in accordance with the logical implications of all that you know. Is it possible to improve the theory in this respect, making allowance within it for the cost of thinking, or would that entail paradox, as I am inclined to believe but unable to demonstrate? (Savage, 1967, p. 308).

1.1.2 Approximation systems

If logic really were informationally trivial, or "tautological", there would be no point in classifying inferences according to their depth. So, in accordance with the traditional idealized view, logical formalisms typically provide no adequate means to measure it. On the other hand, logic is informationally trivial only for ideal agents with unlimited resources, and finding a logical characterization of the depth of logical inference is an interesting challenge both from the theoretical and practical viewpoint.

Prompted by this problem, we endorse a more demanding view of what a fully-fledged logical system should consist of (D'Agostino *et al.*, 2013; Gabbay, 2014; D'Agostino, 2015), that is:

- a mathematically defined consequence relation $\vdash$ (the set of pairs $\langle \mathfrak{A}, \mathfrak{B} \rangle$ such that $\mathfrak{A} \vdash \mathfrak{B}$), representing the "ideal logic", where $\mathfrak{A}, \mathfrak{B}$ are suitable structures of formulae from a logical language $\mathcal{L}$;

- an algorithmic proof systems for generating $\vdash$;

- a definition of how the ideal logic $\vdash$ can be approximated in practice by realistic (not logically omniscient) reasoning agents.

The last component is what we call an "approximation system".

Let $\mathcal{L}$ be an arbitrary propositional language with a finite number of logical operators of fixed finite arity. An *approximation system* for an ideal logic L is a triple

$$\mathcal{A} = \langle \mathbb{P}, \preceq, \{\vdash_\alpha\}_{\alpha \in \mathbb{P}} \rangle$$

where $\preceq$ is a partial ordering of $\mathbb{P}$ such that $(\mathbb{P}, \preceq)$ is a directed set,[5] called the *parameter set*, and $\{\vdash_\alpha\}_{\alpha \in \mathbb{P}}$ is a family of relations between suitable structures of $\mathcal{L}$-formulae such that:

- $\alpha \prec \beta$ implies $\vdash_\alpha \subset \vdash_\beta$,

- for each $\alpha \in \mathbb{P}$, $\vdash_\alpha$ is decidable in polynomial time,

- $\bigcup_{\alpha \in \mathbb{P}} \vdash_\alpha = L$.

Each relation $\vdash_\alpha$ is an *approximation* of L. Clearly, approximation systems are of practical and theoretical interest whenever the ideal logic L is known or conjectured to be intractable. In the context of this book we shall be concerned with approximation systems where the ideal logic is *classical propositional logic* over some arbitrary propositional language $\mathcal{L}$, and the parameter set $\mathbb{P}$ is the set $\mathbb{N}$ of natural numbers totally ordered by $\leq$.

Observe that there is no *a priori* reason that a set of logical operators that is functionally complete for Boolean logic be functionally complete also for all its approximations. So the specification of the logical operators of $\mathcal{L}$ is essential.

To simplify our treatment we shall assume that $\mathcal{L}$ includes at least the four standard Boolean operators associated with the usual Boolean functions. We shall use p, q, r, possibly with subscripts, as metalinguistic variables for atomic $\mathcal{L}$-formulae, A, B, C, etc., possibly with subscripts, for arbitrary $\mathcal{L}$-formulae, and Γ, Δ, Λ, possibly with subscripts, for sets of $\mathcal{L}$-formulae.

When defining an approximation system for Boolean logic we further require that the family $\{\vdash_\alpha\}_{\alpha \in \mathbb{P}}$ satisfies the following conditions, for all A, B, Γ, Δ:

- Local Reflexivity: $A \vdash_\alpha A$ for every $\alpha \in \mathbb{P}$;

- Local Monotonicity: $\Gamma \vdash_\alpha A \Longrightarrow \Gamma, \Delta \vdash_\alpha A$ for every $\alpha \in \mathbb{P}$;

---
[5]This means that every finite subset of $\mathbb{P}$ has an upper bound in $\mathbb{P}$.

1.1. AN INFORMATIONAL VIEW OF CLASSICAL LOGIC

- Global Transitivity: $\Gamma \vdash_\alpha A$ and $\Gamma, A \vdash_\beta B \implies \Gamma \vdash_\gamma B$ for some $\gamma \in \mathbb{P}$ such that γ is an upper bound of α and β;

- Global Substitution Invariance: $\Gamma \vdash_\alpha A \implies \delta(\Gamma) \vdash_\beta \delta(A)$ for some $\beta \succeq \alpha$, where $\delta(A)$ is the result of uniformly replacing atomic formulae of A with arbitrary formulae and $\delta(\Gamma) = \{\delta(B) \mid B \in \Gamma\}$.

1.1.3 Actual vs virtual information

Let us go back to the fundamental question (1.1). By saying that an agent "holds" a piece of information we mean that this information is *actually* accessible to the agent, namely accessible in practice as opposed to being only accessible in principle. After all, at least in its ordinary sense, the information that can be sensibly ascribed to an agent is related to the agent's disposition to answer questions and operate on the basis of the answers. This is what in the Prequel we called *actual information*.

Unlike potential information, actual information is not closed under unrestricted logical consequence. However, it can safely be assumed to be closed under some "easy" inference patterns. Going back to the sudoku example in the Prequel, although there is an algorithm to answer all the questions about the content of each empty cell for a sudoku puzzle of any finite size, we cannot sensibly assume that any realistic rational agent is always capable of answering such question with certainty. The mastery of a decision procedure is by no means sufficient. Such a claim would require a decision procedure such that the cost of obtaining the correct answer is deemed "reasonable" compared to the cost of reviewing the information that is already available. On the other hand for a certain class of sudoku puzzles, that are usually classified as "easy", such undemanding procedures are indeed available and are based on very simple inference steps that can be performed virtually by any agent at a "reasonable" cost, for example that of checking regions and applying the single candidate principle whenever possible.

By contrast, when applying the consensus principle we are processing information that we do not actually possess, and is not in any sense contained in the data. This is what in the Prequel we called *virtual information*. The need for introducing and processing virtual information clearly increases the complexity of the reasoning task, especially when several nested uses of the consensus principle are required, until it becomes unfeasible. As a consequence, whether or not an agent a "holds", in the sense clarified above, a piece of information that requires such complex processing crucially depends on the resources available to the agent. Hence, we cannot assume that actual information is closed under reasoning patterns, such as the consensus principle, that make use of virtual information.

The use of virtual information is essential in several important inference schemes. Consider, for example, a very familiar one, known as "reasoning by cases"

$$\frac{A \vee B \\ A \rightarrow C \\ B \rightarrow C}{C.}$$

To justify this inference scheme we typically reason as follows. Given that $A \vee B$ is true, at least one of A and B must be true. *Suppose* that A is true, then by the second premise C must be true. Alternatively, *suppose* that B is true, then by the third premise C is true again. Hence, C is true, given the premises, no matter which of the two alternative suppositions turns out to be true. This argument cannot be carried out in terms of actual information only. The suppositions constitute information that is not even implicitly contained in the current information state, which is introduced temporarily to reach the conclusion and then disappears when the task has been accomplished. This use of virtual information is epitomized in the well-known natural deduction rule for eliminating disjunction:

$$\frac{A \vee B \quad \begin{matrix}[A] \\ \vdots \\ C\end{matrix} \quad \begin{matrix}[B] \\ \vdots \\ C\end{matrix}}{C}$$

where the conclusion C no longer depend on the suppositions A and B (in square brackets) that are "discharged" by the rule application.

Virtual information plays a crucial role also in "hypothetical reasoning", which is epitomized by the conditional introduction rule of natural deduc-

1.1. AN INFORMATIONAL VIEW OF CLASSICAL LOGIC

tion:
$$\frac{\begin{array}{c}[A]\\ \vdots\\ B\end{array}}{A \to B}$$

Here A is a piece of virtual information that is removed ("discharged") once the conclusion has been obtained.

Other natural deduction rules that require the discharge of virtual assumptions formalize the arguments based on *reductio ad absurdum*:

$$\frac{\begin{array}{c}[A]\\ \vdots\\ \lambda\end{array}}{\neg A} \qquad \frac{\begin{array}{c}[\neg A]\\ \vdots\\ \lambda\end{array}}{A}$$

where λ is the "falsum constant" logically equivalent to any contradiction. Both schemes are classically valid, but only the first one is valid intuitionistically.

Finally, another rule that involves the use of virtual information represents what we have called the "consensus principle" in the Prequel:

$$\frac{\begin{array}{cc}[A] & [\neg A]\\ \vdots & \vdots\\ C & C\end{array}}{C}$$

which is sometimes called "classical dilemma" and is used as an alternative to classical *reductio*. None of the other natural deduction rules involves virtual information.[6]

Such use of virtual information is part of our common reasoning practice and is not too problematic in the most simple cases, but the inference steps that require it appear to be *essentially different* from, and *more demanding* than, those which do not. It is the unbounded use of such steps that makes logical inference a cognitively and computationally hard task. This claim is

---

[6] See (Tennant, 1990; Negri & von Plato, 2001; Mancosu *et al.*, 2021) for excellent expositions

substantiated by the fact, recalled above, that even the pure implication fragment of intuitionistic logic, which is characterized by adding the discharge rule of conditional introduction to *modus ponens* (which is not a discharge rule), PSPACE-complete.

1.1.4 The depth-bounded project

To summarize, even in the case of propositional logic, the conclusion of a difficult inference may carry implicit information that is not *held*, in the sense of being possessed in practice, by every agent who holds the information carried by the premises, and so does not constitute actual information in our sense. There is not (and probably there will never be, according to the dominant conjecture that $P \neq NP$) a feasible procedure for extracting all this implicit information and so there must be a sense of information according to which the class of all valid inferences of classical logic, even at the propositional level, is not informationally trivial. An adequate formalism for classical propositional logic should provide us with the means to measure the *degree of depth* at which implicit information is hidden in the premises and so account for the cost of digging it up. We need an approximation system in which the various tractable subsystems reflect the logical depth of inferences, not merely their length compared to the input length, for the latter method either would be inapplicable or would probably lead to paradoxes. Tractability of depth-bounded subsystems should by no means be involved in their definition, but should emerge as a side result of a sensible measure of depth that applies to *single inferences* depending on their logical structure.

Observe that bounding the use of the discharge rules in Gentzen-style natural deduction would not bring us any closer to a suitable solution of the approximation problem, since (i) these rules are an essential ingredient of the meaning of the logical operators and limiting their use would consist in a change of meaning from one layer to another, (ii) in the basic layer it would be impossible to justify very basic inference principles such as "the single candidate principle" (aka "Disjunctive syllogism") which are paradigmatic of "easy" inference steps.

The overall blueprint of the depth-bounded research program can be traced back to Mondadori (1988a,b, 1989b, 1995); D'Agostino (1990, 1999,

1.1. AN INFORMATIONAL VIEW OF CLASSICAL LOGIC 13

2005, 2010); D'Agostino & Mondadori (1994); Finger (2004a,b); Finger & Gabbay (2006); D'Agostino & Floridi (2009), and can be outlined as follows:

- The classical meaning of the Boolean operators is somehow *overdetermined* by the standard truth-functional semantics to the effect that knowledge of this meaning is not fully manifestable in practice, since the class of inferences that are justified solely on the basis of this meaning are computationally hard.

- Boolean inferences are better construed as arising from the interplay between a weaker "informational" meaning of the logical operators and informational versions of the principles of *Bivalence* and *Non-Contradiction*, which are purely "structural"[7] and do not concern this meaning. The former principle concerns the processing of what we have called "virtual information". The second detects a situation in which part of our information is patently unreliable. In Wittgenstein's words: "Contradiction is to be regarded, not as a catastrophe, but as a wall indicating that we can't go on here" (*Zettel*, §687).

- The separation between the inferential role played by the informational meaning of the logical operators and the inferential role played by the structural principles naturally prompts for the definition of *depth-bounded* approximations in which nested applications of the informational version of the Principle of Bivalence are bounded above by a fixed natural number. The resulting approximation system yields a purely informational view of classical propositional logic and may prove to be an adequate formalism for the investigations of real resource-bounded reasoning agents.

Similar ideas were implicit in the work of Gunnar Stålmarck (1992) and have been independently pursued (Sheeran & Stålmarck, 2000; Björk, 2003, 2005, 2009) in a purely algorithmic framework with more practical motivations, leading to efficient and widely used techniques for software verification (see the Conclusion for further details).

---
[7]We borrow this term from proof-theory to mean that they do not make reference to any logical operator, but apply to any sentence regardless of its form.

The basic layer of this approximation system, the inferences that can be classified as "easy" or "shallow", play a special role, and their semantical and proof-theoretical characterization will be our main task in the next few sections.

1.2 Informational semantics

1.2.1 The need for an informational semantics

Our discussion suggests a plausible necessary condition for a class of inferences to be really "informationally trivial" or tautologous:

(1.2) A class of inferences are informationally trivial *only if* any rational agent can be assumed to be able to recognize *in practice* that the information carried by the conclusion is implicitly contained in the information carried by the premises at a cost that is "reasonable" compared to the cost of grasping premises and conclusion.

In other words:

(1.3) A class of inferences are informationally trivial *only if* they admit of a feasible decision procedure.

Only for such classes could we sensibly claim that a positive answer to the fundamental question (1.1) can be justified. But, which procedures should be considered "feasible"? The identification of such procedures with those that run in polynomial time is itself an idealization that is often put in question. One answer is that our aim in this book is to define a hierarchy of subsystems of classical propositional logic that reflect increasing degrees of idealization and decreasing degrees of feasibility.[8]

The existence of feasible decision procedure is a necessary, but by no means, a sufficient condition for characterizing the basic layer of the informationally trivial inferences. The latter are informationally trivial because they are "analytic" in the sense that their validity depends solely on the meaning of some words, namely the "logical words" (connectives and quantifiers). Analytic inferences are sound by virtue of mere semantic analysis and convey no genuinely new information (see Chapter 4 below). That

---

[8] A more sophisticated answer will be provided in the following chapter (Remark 2.1.1).

1.2. INFORMATIONAL SEMANTICS

the conclusion follows from the premises is a purely linguistic truth that we recognize by virtue only of the meaning of the logical operators occurring in the sentences, and so depends solely on the conventions governing our use of language. But, what is the meaning of the logical operators?

Classical logicians have a straightforward answer to this question: the time-honoured truth tables fix the meaning of each logical operator by fixing the conditions under which a sentence containing it as the main operator is true or false in terms of the truth or falsity of its immediate components. These are conditions that obtain or do not obtain independently of the information that is actually available to us, i.e. our actual information. For example, a disjunction $A \vee B$ is true if and only if either A is true or B is true or they are both true. We cannot express this condition in terms of actual information. The condition "we actually possess the information that $A \vee B$ is true if and only if, for at least one of the two disjuncts, we actually possess the information that it is true" is clearly unsound, since it is quite common that we are informed that a certain disjunction is true without being informed about the truth-value of either disjunct. Indeed, it is mostly under these circumstances that disjunctions are practically useful.

As far as logic is concerned, truth and falsity are taken as primary semantic notions and what counts is only that they obey the classical principles of *bivalence* and *non-contradiction*. According to the first, each sentence, in a given state of affairs, is either determinately true or determinately false quite independently of our capability of recognizing which is the case. According to the second, no sentence can be at the same time true and false. Both principles can be concisely expressed by assuming that a sentence is false if and only if it is not true. This way of fixing the meaning of the logical operators is perfectly in tune with the classical, information-transcendent, notions of truth and falsity and with the traditional view of logical inference as a truth-transmission device; but it is at odds with the equally important view of logical inference as an information-processing device. There is a fundamental *mismatch* between the central semantic notions in terms of which the meaning of the classical operators is defined and the informational nature of the fundamental question (1.1). If the meaning of the classical operator is taken to be their classical, truth-table meaning, all classically valid inferences turn out to be analytic and, therefore, informationally trivial.

In the previous section, we argued that the traditional claim that classical propositional logic is uninformative is untenable for non-ideal agents because the implicit information that it ascribes to them by virtue of the meaning of the logical operators cannot, in general, be feasibly extracted from the information they possess explicitly, and so is not fully manifestable. Hence, it seems that classical semantics assigns the logical operators a meaning that is too strong to justify the claim with respect to a more realistic, fully manifestable, notion of implicit information.

This raises the following:

Problem 1. *Can we fix a* weaker *meaning of the classical operators, we may call it their "informational meaning", in such a way that all inferences that are analytic, i.e., are justified solely in terms of this meaning, are also informationally trivial?*

This, by (1.3), amounts to saying that inferences that are analytic according to the informational meaning of the logical operators, whatever this may turn out to be, should turn out to admit of a feasible decision procedure. According to our analysis, an informational semantics apt to address Problem 1 should fix the meaning of the logical operators in terms of central semantic notions of an informational nature, more specifically in terms of what we have called "actual information".

The obvious move is to replace classical truth and falsity with *informational truth* and *informational falsity*. By saying that a sentence A is informationally true (respectively false) for an agent a — no matter whether human, artificial or hybrid — we mean that, when prompted with the query 'A?', a is in a position of *feasibly* recognizing whether the conditions for answering "yes" (respectively "no") obtain. In our terminology (p. 9), the agent *holds* the information that A is true (respectively false). This is, however, just a way of speaking and we may alternatively speak of the agent's disposition to assent or dissent from a sentence. The way in which such assent and dissent are related to underlying information-transcendent notions of truth and falsity, whatever their nature may be, or even whether such notions are practically or theoretically useful from a logical point of view, is not our concern in the context of this book. Important as this philosophical issue may be, there is no need to elaborate on it for our purposes, and we shall treat informational truth and falsity as *primitive* values that apply to sentences.

1.2. INFORMATIONAL SEMANTICS

Unlike their classical counterparts, informational truth and falsity do not obey the Principle of Bivalence (or, better, its informational version, that we might call "principle of omniscience"): we cannot assume that for every sentence A either we hold the information that A is the case or we hold the information that A is not the case, unless we master a feasible decision procedure. Otherwise, A may well be informationally *indeterminate*. On the other hand, we may safely assume that they satisfy the informational version of the Principle of Non-Contradiction: no agent can *hold*, in the sense specified above, both the information that A is the case and, at the same time, the information that A is not the case, as this would be deemed to be equivalent to possessing no information about A.[9]

Note that the old familiar truth tables for $\wedge$, $\vee$ and $\neg$ are still intuitively sound under this informational reinterpretation of the central semantic notions. For example, if we hold the information that A is true and the information that B is true, we thereby hold the information that $A \wedge B$ is true. If we hold the information that A is true or the information that B is true, we thereby hold the information that $A \vee B$ is true, etc. However, they are no longer exhaustive: they do not tell us what happens when one or all of the immediate components of a complex sentence are indeterminate. Our task is now to redefine in a satisfactory way the meaning of the classical logical operators in terms of the new primary semantic notions of informational truth and informational falsity — or, if you wish, to characterize that *part* of their meaning that can be expressed in these terms. This is what we shall take, in the context of this book, as the *informational meaning of the logical operators*.

Let us re-interpret the values 1 and 0 to denote, respectively, informational truth and falsity. When a sentence takes neither of these two defined values, we say that it is *informationally indeterminate*. We call *partial valuation* any partial mapping from the set of well-formed formulae of a standard propositional language $\mathcal{L}$ to the the set $\{1, 0\}$ of the two determinate

---

[9]This appears quite plausible taking into account that the contradictory pieces of information are supposed, by definition, to be both practically available to the agent, so that the contradiction can be feasibly recognized and this kind of epistemic inconsistency may well be regarded as a sort of uncertainty that makes A informationally indeterminate. Of course, philosophers and logicians may find application contexts in which even this minimal requirement could be relaxed.

Figure 1.1: Information ordering of **3**.

informational values. It is sometimes technically convenient to treat informational indeterminacy as a third value that we denote by "$\bot$".[10] In the sequel, the set of the three values will be denoted by **3**. The *information ordering* $\preceq$ on **3** is illustrated in Figure 1.1, that is, $x \preceq y$ ("x is less defined than, or equal to, y") if, and only if, $x = \bot$ or $x = y$ for $x, y \in$ **3**.

So, it might seem that we only need to conservatively extend the classical truth tables with new entries to accommodate the third value $\bot$. More precisely, for every n-ary Boolean operator $\star$, whose classical meaning is fixed by a classical truth-function $f_\star$, we may deem suitable for specifying its informational meaning as given by some function $\hat{f}_\star$ ($\{0, 1, \bot\}^n \to \{0, 1, \bot\}$) satisfying:

(1.4) $\quad \hat{f}_\star(x_1, \ldots, x_n) = f_\star(x_1, \ldots, x_n)$, whenever $x_1, \ldots, x_n \in \{0, 1\}$.

Given our interpretation of the third value $\bot$ as informational indeterminacy, a reasonable requirement would be also that our logical operators are *monotonic* in the following sense:

(1.5) $\quad x_1 \preceq y_1, \ldots, x_n \preceq y_n \implies \hat{f}_\star(x_1, \ldots, x_n) \preceq \hat{f}_\star(y_1, \ldots, y_n)$

1.2.2 Informational vs 3-valued semantics

Let us restrict our attention to the logical operators $\land, \lor, \neg$. Under the requirements (1.4) and (1.5), the tables of Kleene's (strong) 3-valued logic (Kleene, 1952, §64), shown in Table 1.1, may appear as natural candidates to represent their informational meaning. A "conditional" operator $\to$ may

---

[10]This choice of notation is unusual in the logic literature, where the symbol denotes the *falsum* constant, but it is customary in the literature on partial orderings (Davey & Priestley, 2002). We instead use $\curlywedge$ for the falsum constant.

1.2. INFORMATIONAL SEMANTICS

∧	1	0	⊥
1	1	0	⊥
0	0	0	0
⊥	⊥	0	⊥

∨	1	0	⊥
1	1	1	1
0	1	0	⊥
⊥	1	⊥	⊥

¬	
1	0
0	1
⊥	⊥

Table 1.1: Kleene's 3-valued tables.

be defined in terms of the other operators in the usual (classical) way:

$$A \to B =_{\text{def}} \neg A \lor B$$

or equivalently

$$A \to B =_{\text{def}} \neg(A \land \neg B)$$

As far as the definite values (1 and 0) are concerned, Kleene's 3-valued semantics satisfies the same *necessary and sufficient* conditions as classical semantics:

C1 $v(\neg A) = 1$ if and only if $v(A) = 0$;

C2 $v(A \land B) = 1$ if and only if $v(A) = 1$ and $v(B) = 1$;

C3 $v(A \lor B) = 1$ if and only if $v(A) = 1$ or $v(B) = 1$;

C4 $v(A \to B) = 1$ if and only if $v(A) = 0$ or $v(B) = 1$;

C5 $v(\neg A) = 0$ if and only if $v(A) = 1$;

C6 $v(A \land B) = 0$ if and only if $v(A) = 0$ or $v(B) = 0$;

C7 $v(A \lor B) = 0$ if and only if $v(A) = 0$ and $v(B) = 0$;

C8 $v(A \to B) = 0$ if and only if $v(A) = 1$ and $v(B) = 0$.

A partial valuation v satisfying Conditions C1–C8 is said to be *saturated*. More specifically, we say that it is *upward saturated*, if it satisfies the sufficient conditions (their "if" parts), and *downward saturated* if it satisfies the necessary conditions (their "only-if" parts). A *Boolean valuation* is a saturated partial valuation that satisfies the additional condition of being *total*, i.e. defined for all sentences.

Remark 1.2.1. *Observe that, (i) for total valuations, conditions C5–C8 are redundant, in that they can be derived from conditions C1–C4; (ii) if a valuation v is upward saturated and is defined for all the atomic sentences of the language, then v is a Boolean valuation; (iii) if a valuation v is downward saturated, it can be embedded into a Boolean valuation and this is the basis of the completeness proof for the well-known method of semantic tableaux (Smullyan, 1968). The last point is worth stressing in that it does not enforce bivalence, since a partial valuation may well be downward saturated without being total.*

As argued above, under the informational reinterpretation of the truth-values 1 and 0, the sufficient conditions for the informational truth and falsity of complex sentences are all intuitively sound. However, bivalence and some of the necessary conditions fail, in particular the only-if parts of Conditions C3 and C6.

Example 1.2.2. *When trying to login to some internet account far too often we are told that: "either the username or the password is incorrect".*[11] *So we hold the information that the disjunction is true, although both its components are indeterminate.*

Example 1.2.3. *If I suffer from fuzzy vision, I may well hold the information that the digit at which the optician is pointing is either a 5 or a 7, although the sentences "it is a 5" and "it is a 7" are both indeterminate. In this case we also hold the information that their conjunction is false, unless the optician is fooling us pointing at some ambiguous digit.*

Example 1.2.4. *Suppose we put two bills of 50 and 100 euros in two separate envelopes and then we shuffle the envelopes so as to loose track of which contains which. If we pick one of them, we certainly hold the information that it contains either a 50-euro bill or a 100-euro bill, while both components of the disjunction are indeterminate. In this case too, we hold the information that the conjunction is false.*

Example 1.2.5. *"The roulette ball will fall on a red or black or green pocket": by background information about the roulette wheel, we hold the*

---

[11]This very perspicuous and familiar example was suggested by Hykel Hosni during a seminar.

1.2. INFORMATIONAL SEMANTICS

information that this disjunction is true even if all its components are indeterminate.

On the other hand, depending on our background information, when the components of a disjunction or of a conjunction are indeterminate, the disjunction and the conjunction may well be indeterminate.

Example 1.2.6. *An urn contains only black and white numbered balls. The following sentences:*

1. *"A white ball will be drawn next",*

2. *"a black ball will be drawn next",*

3. *"an odd ball will be drawn next",*

4. *"an even ball will be drawn next"*

are all indeterminate. On the other hand, given our background information on the composition of the urn,

- *the disjunction of 1 and 2 is informationally true*

- *the conjunction of 1 and 2 is informationally false*

- *the disjunction of 3 and 4 is informationally true*

- *the conjunction of 3 and 4 is informationally false*

- *all the other conjunctions and disjunction of the four sentences above are informationally indeterminate.*

These examples are representative of a pervasive feature of our disposition to assent to disjunctions and dissent from conjunctions, which may rely on external sources (Example 1.2.2) or be essentially theory-laden, depending on all sorts of background information (Examples 1.2.3–1.2.6). The reader can easily think of many other examples in a wide variety of situations. Their distinguishing feature is our disposition to assent to a disjunction or dissent from a conjunction without possessing any definite information about their immediate components. Indeed, this is typical of our ordinary usage of disjunctions and conjunctions in most practical contexts.

Moreover, as far as classical propositional logic is concerned, the only-if direction of C4 also fails, as a result of the classical translation of the conditional $A \to B$ as $\neg A \vee B$ or $\neg(A \wedge \neg B)$. For these reasons, Kleene's 3-valued tables are not apt to capture the informational meaning of the logical operators $\vee$ and $\wedge$, namely that part of their meaning that can be specified solely in terms of informational truth and informational falsity.

1.2.3 Informational vs intuitionistic semantics

It may seem that another plausible candidate for a suitable informational semantics may be found by looking at intuitionistic logic, where the meaning of the logical operators may indeed be explained (with some difficulty) in terms of a notion of truth as provability or verifiability that is not information-transcendent.[12] Indeed, the well-known Kripke semantics defines logical consequence as truth-preserving over information states that are intuitively described as "points in time (or 'evidential situations'), at which we may have various pieces of information" (Kripke, 1965, p. 100) and the meaning of the logical operators is fixed in terms of such information states and their potential extensions. However, under closer scrutiny, this hypothesis is also to be discarded.

First, since we are looking for the informational meaning of the *classical* operators, an adequate solution should preserve their classical meaning as much as possible, i.e., whatever in Conditions C1–C8 (p. 19) turns out to be still intuitively sound under the informational reinterpretation of the truth-values in terms of informational truth and falsity. This includes, all their if-parts as well as their only-if parts except for C3, C4 and C6. However, while all the if-parts are intuitionistically sound and the only-if parts of C4 and C6 fail, as they should under the informational interpretation, the only-if parts of C5 and C8 are not satisfied. Equating the intuitionistic falsity of a sentence A with the intuitionistic truth of its negation, the only-if part of C5 fails because of the intuitionistic failure of the double negation law. The only-if part of C8 fails because, unlike the falsity of a classical conditional, the falsity of an intuitionistic conditional cannot be expressed in terms of

---

[12]These issues, and all the subtleties that they involve, have been thoroughly discussed in the logical literature, especially in the writings of Michael Dummett; the reader is referred to Dummett (1991b) for an overall picture.

1.2. INFORMATIONAL SEMANTICS

the actual information possessed by an agent in a given state.

Second, the complexity results mentioned in the previous section imply that shifting from the classical meaning of the logical operators to their intuitionistic meaning is no solution to Problem 1, since intuitionistically valid inferences most probably do not admit of a feasible decision procedure.

Third, in Kripke semantics intuitionistic disjunction satisfies the only-if part of C3, while we have found good reasons for rejecting it on the basis of our notions of informational truth and falsity. As Michael Dummett put it:

> I may be entitled to assert "A or B" because I was reliably so informed by someone in a position to know, but if he did not choose to tell me which alternative held good, I could not apply an or-introduction rule to arrive at that conclusion. [...] Hardy may simply not have been able to hear whether Nelson said "Kismet hardy" or "Kiss me Hardy", though he heard him say one or the other: once we have the concept of disjunction, our perceptions themselves may assume an irremediably disjunctive form (Dummett, 1991b, pp. 266–267) [...]
>
> Unlike mathematical information, empirical information decays at two stages: in the process of acquisition, and in the course of retention and transmission. An attendant directing theatre-goers to different entrances according to the colours of their tickets might even register that a ticket was yellow or green, without registering which it was, if holders of tickets of either colours were to use the same entrance; even our observations are incomplete, in the sense that we do not and cannot take in every detail of what is in our sensory fields. That information decays yet further in memory and in the process of being communicated is evident. In mathematics, any effective procedure remains eternally available to be executed; in the world of our experience, the opportunity for inspection and verification is fleeting (Dummett, 1991b, pp. 277–278).

Note the analogy between Dummett's examples and our Examples 1.2.2–1.2.4. Note also that Example 1.2.4 is particularly interesting in that it can be argued that the disjunction has been obtained by means of an or-introduction, except that the information about which envelope contains which bill has decayed during the process of shuffling the envelopes.[13]

Beth's semantics for intuitionistic logic[14] seems to offer a more natural

---

[13]This kind of information decay is exploited in the well-known three card trick.
[14]For an exposition, see (Dummett, 1977).

account of the truth of disjunctive statements at an information state: a disjunction $A \vee B$ may be verified at the actual state even if neither disjunct is, provided we have the means of recognizing that necessarily one or the other (or both) will eventually be verified at some time in the future, that it to say, that the actual information state will necessarily evolve into one in which (at least) one of the two disjuncts is verified, e.g., for $n > 1$, "n is either prime or composite" is verified at the current information state even if the latter does not actually contain the information of which alternative obtains, for we have an algorithm to decide this in a finite number of steps. Again, this meaning cannot be expressed in terms of the actual information held by an agent (in the given state). Moreover, at least outside constructive mathematics, we may well be in a position to assert a disjunction $A \vee B$ even if we have no *guarantee* that we will eventually reach an information state that enables us to establish the truth of one of the two disjuncts. Examples 1.2.2–1.2.4 point to this possibility.

1.3 Constraint-based semantics

In this section we propose a solution to Problem 1 by outlining a novel informational semantics for the logical operators such that the class of inferences that are characterized by this semantics can be feasibly (and easily) recognized as valid. This will be the basic layer of our approximation system for classical propositional logic.

1.3.1 The informational meaning of the logical operators

Our analysis in the previous sections suggests that we should fix the informational meaning of the logical operators solely in terms of the *actual* information possessed by an agent and characterize the inferences that are truly informationally trivial as exactly those that can be justified by this meaning, without any use of virtual information. This is what we shall do in the rest of this chapter. Next, in Chapter 2, we shall gradually re-introduce virtual information by bounding its use, so as to indefinitely approximate full classical propositional logic.

We have seen (pp. 19 ff) that the if-parts of Conditions C1–C8 are all sound under the reinterpretation of 1 and 0 in terms of informational truth

1.3. CONSTRAINT-BASED SEMANTICS

and falsity. The only-if parts that fail are exactly those that are expressed in a disjunctive form, namely C3, C4 and C6. Such necessary conditions cannot be interpreted in terms of what we called "actual information", information that is practically available to the agent in its current information state. Thus, admissible partial valuations of the $\mathcal{L}$ formulae cannot be specified by means of necessary and sufficient conditions such as C1–C8. This brings us to our next problem:

Problem 2. *Can we fix the informational meaning of the logical operators solely in terms of actual information?*

1.3.2 Meaning via negative constraints

Given the failure of the necessary conditions C3, C4 and C6, we shall take a different route according to which defining the informational meaning of the logical operators consists in providing agents with explicit and manifestable information about the rules of a language game. This can be done by means of *negative constraints*, i.e., by specifying which partial valuations are *inadmissible*. In this approach, the meaning of a logical operator is fixed by specifying which uses of it are *not* allowed.

Definition 1.3.1. *The* immediate subformulae *of a non-atomic $\mathcal{L}$-formula are defined as follows:*

- *A is the only immediate subformula of $\neg A$;*

- *the immediate subformulae of $\star(A_1, \ldots, A_n)$ are $A_1, \ldots, A_n$ for any n-ary operator $\star$.*

Definition 1.3.2. *We say that B is a* subformula *of A if there is a sequence $A_0, \ldots, A_n$ such that (i) $A_0 = B$, (ii) A_{i+1} is an immediate subformula of A_i, (iii) $A_n = A$.*

Note that this definition implies that every formula A is a subformula of itself (when $n = 0$).

Definition 1.3.3. *We say that B is a* proper subformula *of A if B is a subformula of A and $B \neq A$.*

Given a logical language $\mathcal{L}$,

Definition 1.3.4. *An $\mathcal{L}$-module of $\mathcal{L}$ is a set $M = \{A_0, A_1, \ldots, A_n\}$ such that A_0 is a non-atomic formula, called* top formula *of M, and $A_1, \ldots, A_n$ are its immediate subformulae, called* secondary formulae *of M. The* main operator *of M is the main operator of its top formula. An $\star$-module is an $\mathcal{L}$-module such that $\star$ is the main operator of its top formula.*

For example: $\{(p \wedge q) \vee (r \to s), p \wedge q, r \to s\}$ is a $\vee$-module.

Definition 1.3.5. *Let v be a partial valuation. A valuation module of v is any mapping $\alpha : M \to \mathbf{3}$ for some $\mathcal{L}$-module M such that $\alpha = v|M$. We shall call* top formula *of α the top formula of M and* main operator *of α the main operator of M.*

The informational meaning of a logical operator $\star$ is fixed by determining which valuation modules are *inadmissible*. Imagine the following dialogue:

> A: Is it true that either Ann or Mary will go to the party?
> B: Yes, however I know that Ann will not go and that Mary will not go either.

The only explanation of B's bizarre answer (barring temporary mental insanity) is that B has no basic knowledge of the meaning of "either...or...". Thus, a valuation module α such that $\alpha(A \vee B) = 1, \alpha(A) = 0$ and $\alpha(B) = 0$ would clearly be inadmissible and this negative constraint can be taken as part of the intended meaning of "$\vee$". Any agent who grasps the meaning of $\vee$, even superficially, can immediately detect that saying that a disjunction is true and both its components are false is not consistent with this meaning. Similarly, a valuation module such that $v(A \wedge B) = 1$, $v(A) = 0$ and $v(B) = \bot$ is inadmissible, as is any of its refinements on B, and this is part of the intended meaning of $\wedge$.

We do not claim that such constraints represent the *full* meaning of the logical operators and that the classical stipulations are, for some reason, not acceptable. They represent only the *part* of this meaning that can be expressed in terms of *actual information* only and can be *easily* manifested by any agent who grasps it even at the most basic level, to rule out inadmissible assignments of informational values; for example by recognizing the inconsistency of receiving the answers "yes" to the question "$A \vee B$?" and "no"

1.3. CONSTRAINT-BASED SEMANTICS

A	¬A
1	1
0	0

A	B	A ∨ B
0	0	1
1	⊥	0
⊥	1	0

A	B	A ∧ B
0	⊥	1
⊥	0	1
1	1	0

A	B	A → B
1	0	1
⊥	1	0
0	⊥	0

Table 1.2: The informational meaning of the logical operators. The rows represent minimal inadmissible valuation modules.

to both the questions "A?" and "B?". On the other hand, any agent who is unable to recognize such patently incorrect uses of ∨ is an agent who has *no degree* of understanding of its accepted meaning. This does not imply, of course, that an agent who grasps this basic part of the meaning of the logical operators is thereby able to perform any logical inference involving them.

The negative constraints for the four standard operators are summarized in Table 1.2, where each line represents a *minimal* inadmissible valuation module, i.e. such that any of its refinements (with respect to ⊑) is also inadmissible. So, every row containing a ⊥ stands for three inadmissible valuation modules. Similar constraints can be derived for any Boolean operator from its truth-table, by identifying the minimal partial valuations such that none of its refinements corresponds to a row. For example, the constraints for *exclusive or* (⊕), *Peirce's arrow* (↓), *Sheffer's stroke* (|) are illustrated in Table 1.3.

A valuation v is admissible when it does not "contain" inadmissible modules. We denote by $V_\star$ the set of all admissible valuation modules such that their main operator is $\star$, i.e. such that their domain is a $\star$-module.

Definition 1.3.6. *A valuation v is* admissible *if $v|M$ is in $V_\star$ for every M such that $\star$ is its main operator.*

In other words, an admissible valuation is one such that the formulae in each $\mathcal{L}$-module M are assigned values according to the most basic "rules of

A	B	$A \oplus B$
0	0	1
1	1	1
1	0	0
0	1	0

A	B	$A \downarrow B$
$\bot$	1	1
1	$\bot$	1
0	0	0

A	B	$A \mid B$
1	1	1
$\bot$	0	0
0	$\bot$	0

Table 1.3: The informational meaning of exclusive or, Peirce's arrow and Sheffer's stroke. Rows represent minimal inadmissible valuations.

the game", namely to the negative meaning constraints for the main operator of its top formula.

We shall denote by $\mathcal{A}$ the domain of all admissible valuations. Admissible valuations are partially ordered by the approximation relation $\sqsubseteq$ defined as follows: $v \sqsubseteq w$ (read "w is a *refinement* of v" or "v is an *approximation* of w") if and only if w agrees with v on all the formulae A for which $v(A) \neq \bot$. Similarly, we say that w is a *strict refinement* of v (or that v is a *strict approximation* of w), written $v \sqsubset w$, if and only if w is a refinement of v and $v \neq w$.

The partial ordering $\sqsubseteq$ is a meet-semilattice with a bottom element, namely the valuation which is undefined for all formulae of the language. It fails to be a lattice because the union of two admissible valuations may be inadmissible.

A partial valuation v can be identified with the set of all pairs $\langle A, x \rangle$ such that $v(A) = x$, for $x \in \{1, 0\}$. Under this representation, A is indeterminate if and only if neither $\langle A, 1 \rangle$ nor $\langle A, 0 \rangle$ belong to this set and the partial ordering $\sqsubseteq$ is simply set inclusion. Each pair $\langle A, x \rangle$ can be thought of as a "piece of information" and the valuation itself as an attempt to put together such pieces of information in a way that is consistent with the intended meaning of the logical operators.

1.3. CONSTRAINT-BASED SEMANTICS

1.3.3 The single candidate principle

Now, the next question is: which inferences can be justified solely by virtue of this informational *approximation* of the classical meaning of the logical operators? Consider, again, a valuation module α such that $\alpha(p \vee q) = 1$ and $v(p) = 0$, while $v(q) = \bot$. We can legitimately say that the value of q is *determined* in α by our understanding of the meaning of $\vee$ based on the constraints specified in Table 1.2. Indeed, there is no admissible *strict* refinement of α such that $\alpha(q) = 0$, since such a refinement would fail to satisfy one of the constraints that define the meaning of "$\vee$". In other words, any assignment of a *defined* value other than 1 would be immediately recognized as inconsistent by any agent that understands $\vee$ via the specified constraints. Under these circumstances, it is straightforward to recognize that the piece of information $\langle B, 1 \rangle$ is *implicitly contained* in v, in particular, in the valuation module α of v with $A \vee B$ as top formula, because the negative constraints on $\vee$ leave us no other *defined* option. This is, in essence, an application of the "single candidate principle" and the logical situation is comparable to what happens in the easy steps of the sudoku game where the digit to be inserted in a given empty cell is *dictated* by the digits already inserted in the cells belonging to the regions in which this cell is contained. In our context, a "region" is simply an $\mathcal{L}$-module and the meaning constraints are the analogues of the sudoku rules.

All the 27 possible valuation modules for the $\mathcal{L}$-module M with top formula $p \vee q$ are listed in Table 1.4.

Some of these valuation modules are inadmissible, namely 8, 15, 17, 20, 23, 26. All the others are admissible. Some of them are such that some of their undefined elements admit of only one possible refinement that is compatible with the meaning constraints: 2, 5, 6, 7, 11, 12, 13, 16, 19, 22, 25. Under these circumstances we say that the module is *unstable on* each of the formulae that admit of only one possible refinement, to convey the idea that these formulae are, as it were, "attracted" towards their only possible defined value. For example, 2 is unstable on both p and q; 5 is unstable only on p and so is 6; 7 is unstable only on $p \vee q$ and so are 13, 16, 19, 22, 25; 11 is unstable only on q and so is 12. A module is *stable* if it is not unstable on any of its formulae. In our example, 1, 2, 3, 4, 9, 10, 14, 15, 17, 18, 21, 26, 27 are all stable.

	p	q	$p \vee q$
1	$\bot$	$\bot$	$\bot$
2	$\bot$	$\bot$	0
3	$\bot$	$\bot$	1
4	$\bot$	0	$\bot$
5	$\bot$	0	0
6	$\bot$	0	1
7	$\bot$	1	$\bot$
8	$\bot$	1	0
9	$\bot$	1	1
10	0	$\bot$	$\bot$
11	0	$\bot$	0
12	0	$\bot$	1
13	0	0	$\bot$
14	0	0	0
15	0	0	1
16	0	1	$\bot$
17	0	1	0
18	0	1	1
19	1	$\bot$	$\bot$
20	1	$\bot$	0
21	1	$\bot$	1
22	1	0	$\bot$
23	1	0	0
24	1	0	1
25	1	1	$\bot$
26	1	1	0
27	1	1	1

Table 1.4: Valuation modules for the $\mathcal{L}$-module $M = \{p, q, p \vee q\}$.

1.3. CONSTRAINT-BASED SEMANTICS   31

Shallow information states

Thus, the single candidate principle (SCP) is the most basic consistency principle by means of which "elementary" (in Sherlock's sense) logical inferences can be justified "analytically", that is, by virtue only of the informational meaning of the logical operators as specified by the negative constraints. The emergence of this most basic kind of deductive reasoning can be figuratively expressed by the following maxim:

| ELEMENTARY INFERENCE = NEGATIVE MEANING CONSTRAINTS + SCP |

Remark 1.3.7. *Note that, in the presence of a finite amount of information (i.e., a finite number of formulae with a defined value), applying the SCP is a task that can be performed efficiently with respect to the available information. Indeed, it requires checking whether a possible strict refinement regarding a specific undefined formula B is admissible, which in turn requires checking only the "neighbourhoods" of B, namely the finitely many $\mathcal{L}$-modules containing B that are not totally undefined. Each module contains a number of formulae which is less than or equal to $a + 1$, where a is the maximum arity of the logical operators in $\mathcal{L}$.*

To summarize, given a valuation module α such that $A = \bot$, we say that a piece of information $\langle A, x \rangle$, with $x \in \{0, 1\}$, is *implicitly contained* in α and write $\alpha \Vdash_{\text{SCP}} \langle A, x \rangle$, if the alternative piece of information $\langle A, |1-x| \rangle$ is *immediately ruled out* by the meaning constraints. In symbols (recalling that $V_\star$ is the set of all admissible valuation modules such that $\star$ is their main operator, see above p. 27):

(SCP) $\alpha \Vdash_{\text{SCP}} \langle A, x \rangle \iff \alpha(A) = \bot$ and $\alpha \cup \{\langle A, 1-x \rangle\} \notin V_\star$,

where $x \in \{0, 1\}$. Observe that a module α is stable if and only if it is closed under SCP, that is: if $\alpha' \Vdash_{\text{SCP}} \langle A, x \rangle$ for some $\alpha' \sqsubseteq \alpha$, then $\langle A, x \rangle \in \alpha$. We say that $v \Vdash_{\text{SCP}} \langle A, x \rangle$ whenever $\alpha \Vdash_{\text{SCP}} \langle A, x \rangle$ for some module α of v.

In this conceptual framework, what is the most basic kind of information state? A *minimal* requirement is that it is closed under the implicit information that immediately stems from the meaning constraints combined with SCP.

Definition 1.3.8. *We say that an admissible valuation v is a* shallow information state *if all its valuation modules are stable (closed under SCP).*

Intuitively, a shallow information state represents the overall information that a reasoner holds, either explicitly or implicitly, on the sole basis of the intended (informational) meaning of the logical operators and of the basic consistency principle expressed by the SCP.

It can be easily verified that a shallow information state is a Boolean valuation if and only if $v(p) \neq \bot$ for every atomic p. So, Boolean valuations can be seen as shallow information states that are closed under a *Principle of Omniscience*, the informational counterpart of the classical Principle of Bivalence:

(PO) For every atomic sentence p, either p is informationally true or p is informationally false.

Example 1.3.9. Suppose v is a shallow information state such that

1. $v(p \vee q) = 1$
2. $v(p) = 0$
3. $v(q \to r) = 1$
4. $v(q \to s) = 1$
5. $v(\neg t \to \neg(r \wedge s)) = 1$
6. $v(t \wedge u) = 0$

We show that $v(u) = 0$. From 1 and 2, by the constraints on $\vee$ it follows that if $v(q) \neq 0$, otherwise v would be inadmissible. Moreover, if $v(q) = \bot$, then $v \Vdash_{\text{SCP}} \langle q, 1 \rangle$, since $v \cup \langle q, 0 \rangle$ would be inadmissible, and so the relevant valuation module

p	q	$p \vee q$
0	$\bot$	1

would be unstable on q. Then, $v(q) \neq \bot$ and, by SCP:

7. $v(q) = 1$.

1.3. CONSTRAINT-BASED SEMANTICS

Next, from 7 and 3 it follows that $v(r) \neq 0$ by the meaning constraints on $\to$. Moreover, if $v(r) = \bot$, then $v \Vdash_{\text{SCP}} \langle r, 1 \rangle$, and the relevant module would be unstable. Hence, by SCP again:

8. $v(r) = 1$.

The remaining steps of the argument are similar and can be summarized as follows:

9. $v(s) = 1$, by 4, 7, the meaning constraints on $\to$ and SCP.

10. $v(r \wedge s) = 1$, by 8, 9, the meaning constraints on $\wedge$ and SCP.

11. $v(\neg(r \wedge s)) = 0$, by 10, the meaning constraints on $\neg$ and SCP.

12. $v(\neg t) = 0$, by 5, 11, the meaning constraints on $\to$ and SCP.

13. $v(t) = 1$, by 12, the meaning constraints on $\neg$ and SCP.

14. $v(u) = 0$, by 6, 13, the meaning constraints on $\wedge$ and SCP.

In what follows, to emphasize the connection with the semantics, we shall make use of *signed formulae* (*S-formulae* for short). A *signed formula* is an expression of the form $T A$ or $F A$, where A is an $\mathcal{L}$-formula, with the intended meaning of "A is informationally true" and "A is informationally false" respectively. Throughout this book (with the exception of Chapter 3, see p. 107), we shall use "$\varphi, \psi, \theta, \ldots$" as variables ranging over S-formulae, "$A, B, C, \ldots$" as variables ranging over unsigned formulae and "$X, Y, Z, \ldots$", as variables ranging over sets of S-formulae and "$\Gamma, \Delta, \Lambda, \ldots$", as variables ranging over sets of unsigned formulae. We shall make use of subscripts to extend our stock of variables. Moreover, in the expression "SA", S stands for either T or F, while in the expression "$\overline{S}A$", $\overline{S}$ stands for T if S is F and for F if S is T. $\overline{S}A$ is called the *conjugate* of SA. In what follows we shall write $S\Gamma$ for $\{SA \mid A \in \Gamma\}$. We shall also write $\overline{\varphi}$ to denote the conjugate of φ.

Let us say that a shallow information state v *satisfies* an S-formula TA if $v(A) = 1$ and an S-formula FA if $v(A) = 0$.

Definition 1.3.10. *For every set X of S-formulae and every S-formula φ, we say that:*

- φ *is a* 0-depth consequence *of X if v satisfies φ for every shallow information state v such that v satisfies all the S-formulae in X.*

- X *is* 0-depth inconsistent *if there is no shallow information state v such that v satisfies all the S-formulae in X.*

We use the symbol "$\vDash_0$" for the 0-depth consequence relation and write "$X \vDash_0 \varphi$" for "φ is a 0-depth consequence of X". We also write "$X \vDash_0$" as shorthand for "$X \vDash_0 \varphi$ for all φ", to mean that X is 0-depth inconsistent. The notions of 0-depth consequence and 0-depth inconsistency can be extended to unsigned formulae by stipulating that an unsigned formula A is a 0-depth consequence of a set Γ of unsigned formulae if and only if $T\Gamma \vDash_0 TA$ and that Γ is 0-depth inconsistent if and only if $T\Gamma$ is 0-depth inconsistent. To simplify the notation we shall use the same symbol $\vDash_0$ for the consequence relation defined on unsigned formulae.

It is not difficult to show that this consequence relation is *a logic in Tarski's sense*, i.e., it satisfies the following conditions:

(Reflexivity) $\quad\quad\quad\quad\quad A \vDash_0 A$

(Monotonicity) $\quad\quad\quad\, \Gamma \vDash_0 A \Longrightarrow \Gamma \cup \Delta \vDash_0 A$

(Cut) $\quad\quad\quad\quad\quad\quad\quad \Gamma \vDash_0 A$ and $\Gamma \cup \{A\} \vDash_0 B \Longrightarrow \Gamma \vDash_0 B$.

(Substitution Invariance) $\quad \Gamma \vDash_0 A \Longrightarrow \delta\Gamma \vDash_0 \delta A$

where δ is an arbitrary substitution of atomic formulae with formulae and $\delta\Gamma$ is short for $\{\delta A | A \in \Gamma\}$.

In Section 1.5.2 it is shown that the 0-depth consequence relation is not a (finite) many-valued logic, that is, it cannot be characterized by any set of finitely valued matrices, a feature that it shares with intuitionistic propositional logic.

Observe that a trivial consequence of Definition 1.3.10 is that the 0-depth logic $\vDash_0$ is *explosive*, that is, $X \vDash_0 \varphi$, for any φ, whenever X is 0-depth inconsistent. However, we shall see that this logic admits of a feasible decision procedure, so that 0-depth inconsistency is easily detected. In the sequel we shall see how explosivity can be avoided by proof-theoretical means.

Finally, an important property of $\vDash_0$ is that, like Kleene's 3-valued logic Kleene (1952) and Belnap's 4-valued logic Belnap Jr. (1976, 1977), it has

1.4. INTELIM SEQUENCES

no "tautologies", i.e., for no formula A, it holds true that $\emptyset \models_0 A$. To see this, it is sufficient to observe that the partial valuation v such that $v(A) = \bot$ for all A is, according to our definition, a shallow information state — we could call it the *null* shallow information state. Hence, there is no formula that is verified by *all* shallow information states; in particular no formula is verified by the null information state.

1.4 Intelim sequences

1.4.1 The emergence of inference rules

We shall now show how some familiar inference rules that do not involve any virtual information, are generated by the meaning constraints and SCP.

By the *logical complexity* of a formula A we mean the number of occurrences of logical operators in it. The logical complexity of a valuation module is the logical complexity of its top formula.

Given a substitution δ, we write $\delta(SA)$ for $S\delta A$, and δX for $\{\delta(\varphi) \mid \varphi \in X\}$.

Definition 1.4.1. *Let*

- α *be a* minimal *(with respect to $\sqsubseteq$) valuation module of complexity 1 such that, for some A, $\alpha(A) = \bot$ and $\alpha \Vdash_{\mathrm{SCP}} \langle A, x \rangle$ (with $x \in \{0,1\}$);*

- $X_\alpha = \{T A \mid \alpha(A) = 1\} \cup \{F A \mid \alpha(A) = 0\}$;

- $\mathcal{R}(\alpha, A) = \{\langle \delta X_\alpha, S_x A \rangle\}_{\delta \in \mathcal{S}}$ *where*

  - $\mathcal{S}$ *is the set of all substitutions of atomic formulae with arbitrary ones,*
  - $S_x = \begin{cases} T & \text{if } x = 1 \\ F & \text{if } x = 0. \end{cases}$

Then,

(a) $\mathcal{R}(\alpha, A)$ *is an* inference rule generated by α;

(b) *if A is the top formula of α, $\mathcal{R}(\alpha, A)$ is an* introduction rule; *otherwise, $\mathcal{R}(\alpha, A)$ is an* elimination rule;[15]

(c) *each pair in $\mathcal{R}(\alpha, A)$, is an* application *of the rule with δX_α as* premises *and $\delta(S_x A)$ as conclusion.*

By virtue of of the above definition we can represent introduction and elimination rules by means of formula-schemes obtained by replacing the atomic formulae occurring in the domain of α with metalinguistic variables in such a way that distinct atomic letters are replaced by distinct metalinguistic variables.

The introduction and elimination rules (from now on *intelim* rules) for the four standard logical operators generated by the respective meaning constraints are shown in Table 1.5 and are expressed in terms of S-formulae. In Table 1.7 we show the intelim rules for signed formulae generated by the meaning constraints in Table 1.3.

For the sake of philosophical analysis, the rules are best presented in this form. The standard reading of signed formulae in the context of classical logic is "A is true" and "A is false" and this appears to be the simplest way of achieving "separation" in the sense of (Bendall, 1978, p. 250) — each rule deals only with one logical operator and a proof should make use only of intelim rules for the operators that occur in the premises or in the conclusion — as well as the stronger form of separation that is embodied in the subformula property. This is also the approach followed by Smullyan (1968) in his presentation of the tableau method. For a well-argued philosophical defense of the use of signed formulae in the proof theory of classical logic the reader is referred to (Bendall, 1978).[16] An alternative, but closely related, way of achieving the same results in terms of ordinary formulae is that of resorting to *multi-conclusion sequents* as the primary components of a deduction system. The connection between multi-conclusion sequents and signed formulae is made apparent when looking at the translation of classical sequent proofs in systems with no structural rules like Kleene's G4 (Kleene, 1967, Chapter 6) into closed semantic tableaux.[17]

---

[15] It can be verified that in this case $\alpha(A) \neq \bot$.

[16] More recently, the use of signed formulae for philosophical purposes has been central in the discussion on "bilateralism" (Smiley, 1996; Rumfit, 2000; Humberstone, 2000; Ferreira, 2008; Gabbay, 2017) as an inferential approach to the meaning of the classical operators.

[17] See (D'Agostino, 1990, Section 2.2) for the details. For another interesting approach,

1.4. INTELIM SEQUENCES

For all practical purposes, however, we may find it convenient to work with *unsigned* formulae, by exploiting the classical meaning of negation. This amounts to simply removing the sign T before a formula and replacing the sign F with the negation sign $\neg$. In this version, the intelim rules are no longer "separated", the subformula property holds only in a weaker form (see p. 53). A version of the intelim rules for unsigned formulae is obtained by "unsigning" the signed formulae as just explained and removing redundant rules, as in Table 1.6. In the unsigned version the two-premise elimination rules correspond to time-honoured principles of inference: *modus ponens*, *modus tollens*, *disjunctive syllogism* and its dual.[18] In the elimination rules, the formula containing the logical operator that is to be eliminated is called *major premise* and the other (if any) is called *minor premise*. The less natural rules, from the point of view of ordinary usage, namely the introduction rules for a true conditional and the elimination rules for a false conditional, are related to the Philonian meaning of this operator which is, however, typical of classical logic. The Philonian conditional, also called "material implication", is defined by $A \to B =_{\text{def}} \neg A \vee B$ or, equivalently, by $A \to B =_{\text{def}} \neg(A \wedge \neg B)$.

1.4.2 Intelim sequences

The application of the intelim rules of Table 1.5 generates sequences of S-formulae, that we call "intelim sequences".

Definition 1.4.2. *Given a set X of S-formulae:*

- *An* intelim sequence *for X is a sequence $\varphi_1, \ldots \varphi_n$ of S-formulae such that, for every $i = 0, \ldots, n$, either $\varphi_i \in X$ or φ_i is the conclusion of the application of an intelim rule to preceding formulae.*

---

in the context of the Curry-Howard correspondence for classical logic, see (Aschieri *et al.*, 2018). On the other hand, without essential extensions of the logical language (signed formulae) or of the intuitive notion of inference rule (multi-conclusion sequent calculus), or other non-standard technical devices most authors are skeptical about the possibility of a genuine inferential semantics for classical logic. One notable exception is (Sandqvist, 2009); see also the analysis in (Makinson, 2013).

[18]Chrysippus (III century B.C.) listed these rules among the fundamental indemonstrable principles of reasoning (*anapodeiktoi*), except that he intended disjunction in its exclusive sense.

$$\frac{FA}{T\neg A}\,F\neg\text{-}\mathcal{I} \qquad \frac{TA}{F\neg A}\,T\neg\text{-}\mathcal{I} \qquad \frac{T\neg A}{FA}\,T\neg\text{-}\mathcal{E} \qquad \frac{F\neg A}{TA}\,F\neg\text{-}\mathcal{E}$$

$$\frac{TA}{TA\vee B}\,T\vee\text{-}\mathcal{I}1 \qquad \frac{TB}{TA\vee B}\,T\vee\text{-}\mathcal{I}2 \qquad \frac{FA\quad FB}{FA\vee B}\,F\vee\text{-}\mathcal{I}$$

$$\frac{TA\vee B\quad FA}{TB}\,T\vee\text{-}\mathcal{E}1 \qquad \frac{TA\vee B\quad FB}{TA}\,T\vee\text{-}\mathcal{E}2 \qquad \frac{FA\vee B}{FA}\,F\vee\text{-}\mathcal{E}1 \qquad \frac{FA\vee B}{FB}\,F\vee\text{-}\mathcal{E}2$$

$$\frac{FA}{FA\wedge B}\,F\wedge\text{-}\mathcal{I}1 \qquad \frac{FB}{FA\wedge B}\,F\wedge\text{-}\mathcal{I}2 \qquad \frac{TA\quad TB}{TA\vee B}\,T\wedge\text{-}\mathcal{I}$$

$$\frac{FA\wedge B\quad TA}{FB}\,F\wedge\text{-}\mathcal{E}1 \qquad \frac{FA\wedge B\quad TB}{FA}\,F\wedge\text{-}\mathcal{E}2 \qquad \frac{TA\wedge B}{TA}\,T\wedge\text{-}\mathcal{E}1 \qquad \frac{TA\wedge B}{TB}\,T\wedge\text{-}\mathcal{E}2$$

$$\frac{FA}{TA\rightarrow B}\,T\rightarrow\text{-}\mathcal{I}1 \qquad \frac{TB}{TA\rightarrow B}\,T\rightarrow\text{-}\mathcal{I}2 \qquad \frac{TA\quad FB}{FA\rightarrow B}\,F\rightarrow\text{-}\mathcal{I}$$

$$\frac{TA\rightarrow B\quad TA}{TB}\,T\rightarrow\text{-}\mathcal{E}1 \qquad \frac{TA\rightarrow B\quad FB}{FA}\,T\rightarrow\text{-}\mathcal{E}2 \qquad \frac{FA\rightarrow B}{TA}\,F\rightarrow\text{-}\mathcal{E}1 \qquad \frac{FA\rightarrow B}{FB}\,F\rightarrow\text{-}\mathcal{E}2$$

Table 1.5: Intelim rules for the four standard Boolean operators.

1.4. INTELIM SEQUENCES

$$\frac{A}{\neg\neg A}\ \neg\neg\text{-}\mathcal{I} \qquad \frac{\neg\neg A}{A}\ \neg\neg\text{-}\mathcal{E}$$

$$\frac{A}{A \vee B}\ \vee\text{-}\mathcal{I}1 \qquad \frac{B}{A \vee B}\ \vee\text{-}\mathcal{I}2 \qquad \frac{\neg A \quad \neg B}{\neg(A \vee B)}\ \neg\vee\text{-}\mathcal{I}$$

$$\frac{A \vee B \quad \neg A}{B}\ \vee\text{-}\mathcal{E}1 \qquad \frac{A \vee B \quad \neg B}{A}\ \vee\text{-}\mathcal{E}2$$

$$\frac{\neg(A \vee B)}{\neg A}\ \neg\vee\text{-}\mathcal{E}1 \qquad \frac{\neg(A \vee B)}{\neg B}\ \neg\vee\text{-}\mathcal{E}2$$

$$\frac{\neg A}{\neg(A \wedge B)}\ \neg\wedge\text{-}\mathcal{I}1 \qquad \frac{\neg B}{\neg(A \wedge B)}\ \neg\wedge\text{-}\mathcal{I}2 \qquad \frac{A \quad B}{A \wedge B}\ \wedge\text{-}\mathcal{I}$$

$$\frac{\neg(A \wedge B) \quad A}{\neg B}\ \neg\wedge\text{-}\mathcal{E}1 \qquad \frac{\neg(A \wedge B) \quad B}{\neg A}\ \neg\wedge\text{-}\mathcal{E}2$$

$$\frac{A \wedge B}{A}\ \wedge\text{-}\mathcal{E}1 \qquad \frac{A \wedge B}{B}\ \wedge\text{-}\mathcal{E}2$$

$$\frac{\neg A}{A \to B}\ \to\text{-}\mathcal{I}1 \qquad \frac{B}{A \to B}\ \to\text{-}\mathcal{I}2 \qquad \frac{A \quad \neg B}{\neg(A \to B)}\ \neg\to\text{-}\mathcal{I}$$

$$\frac{A \to B \quad A}{B}\ \to\text{-}\mathcal{E}1 \qquad \frac{A \to B \quad \neg B}{\neg A}\ \to\text{-}\mathcal{E}2$$

$$\frac{\neg(A \to B)}{A}\ \neg\to\text{-}\mathcal{E}1 \qquad \frac{\neg(A \to B)}{\neg B}\ \neg\to\text{-}\mathcal{E}2$$

Table 1.6: Intelim rules for the four standard Boolean operators (unsigned version).

$$\frac{T\,A \quad T\,B}{T\,A \oplus B}\,T\oplus\text{-}\mathcal{I}1 \qquad \frac{F\,A \quad T\,B}{T\,A \oplus B}\,T\oplus\text{-}\mathcal{I}2 \qquad \frac{T\,A \quad T\,B}{F\,A \oplus B}\,F\oplus\text{-}\mathcal{I}1 \qquad \frac{F\,A \quad F\,B}{F\,A \oplus B}\,F\oplus\text{-}\mathcal{I}2$$

$$\frac{T\,A \oplus B \quad T\,A}{F\,B}\,T\oplus\text{-}\mathcal{E}1 \qquad \frac{T\,A \oplus B \quad T\,B}{F\,A}\,T\oplus\text{-}\mathcal{E}2 \qquad \frac{T\,A \oplus B \quad F\,A}{T\,B}\,T\oplus\text{-}\mathcal{E}3 \qquad \frac{T\,A \oplus B \quad F\,B}{T\,A}\,T\oplus\text{-}\mathcal{E}4$$

$$\frac{F\,A \oplus B \quad T\,A}{T\,B}\,F\oplus\text{-}\mathcal{E}1 \qquad \frac{F\,A \oplus B \quad T\,B}{T\,A}\,F\oplus\text{-}\mathcal{E}2 \qquad \frac{F\,A \oplus B \quad F\,A}{F\,B}\,F\oplus\text{-}\mathcal{E}3 \qquad \frac{F\,A \oplus B \quad F\,B}{F\,A}\,F\oplus\text{-}\mathcal{E}4$$

$$\frac{T\,A}{F\,A \downarrow B}\,T\downarrow\text{-}\mathcal{I}1 \qquad \frac{T\,B}{F\,A \downarrow B}\,T\downarrow\text{-}\mathcal{I}2 \qquad \frac{F\,A \quad F\,B}{T\,A \downarrow B}\,T\downarrow\text{-}\mathcal{I}$$

$$\frac{F\,A \downarrow B}{F\,A \quad T\,B}\,F\downarrow\text{-}\mathcal{E}1 \qquad \frac{F\,A \downarrow B}{F\,B \quad T\,A}\,F\downarrow\text{-}\mathcal{E}2 \qquad \frac{T\,A \downarrow B}{F\,A}\,T\downarrow\text{-}\mathcal{E}1 \qquad \frac{T\,A \downarrow B}{F\,B}\,T\downarrow\text{-}\mathcal{E}2$$

$$\frac{F\,A}{T\,A\mid B}\,T\mid\text{-}\mathcal{I}1 \qquad \frac{F\,B}{T\,A\mid B}\,T\mid\text{-}\mathcal{I}2 \qquad \frac{T\,A \quad T\,B}{F\,A\mid B}\,F\mid\text{-}\mathcal{I}$$

$$\frac{T\,A\mid B \quad T\,A}{F\,B}\,T\mid\text{-}\mathcal{E}1 \qquad \frac{T\,A\mid B \quad T\,B}{F\,A}\,T\mid\text{-}\mathcal{E}2 \qquad \frac{F\,A\mid B}{T\,A}\,F\mid\text{-}\mathcal{E}1 \qquad \frac{F\,A\mid B}{T\,B}\,F\mid\text{-}\mathcal{E}2$$

Table 1.7: Intelim rules for *exclusive or* ($\oplus$), *Peirce's arrow* ($\downarrow$), *Sheffer's stroke* ($\mid$).

1.4. INTELIM SEQUENCES

- *An intelim sequence is* closed *when it contains both T A and F A for some A. It is* atomically closed *if it contains both T p and F p for some atomic formula p.*

- *An intelim* refutation *of X is a closed intelim sequence for X.*

- *An intelim* proof *of φ from X is an intelim sequence for X such that φ is the last S-formula in the sequence.*

- *A set X of S-formulae is intelim* refutable *if there is a closed intelim sequence for X.*

- *An S-formula φ is intelim* deducible *from X if there is an intelim proof of φ from X.*

- *A set X of S-formulae is* intelim inconsistent *if X is intelim refutable.*

We can extend the notions of intelim deducibility and refutability to unsigned formulae, as we did for 0-depth consequence and 0-depth inconsistency, by stipulating that an unsigned formula A is intelim deducible from a set Γ of unsigned formulae if $T A$ is intelim deducible from $T \Gamma$ and that a set Γ of unsigned formulae is intelim refutable if $T \Gamma$ is intelim refutable. Alternatively, we can re-define the notion of intelim sequence and all the related notions directly in terms of unsigned formulae, using the rules in Table 1.7.

In Figure 1.2 we show simple examples of intelim sequences using, respectively, the intelim rules for signed formulae and their version for unsigned formulae. To avoid notational proliferation, we use the same symbol "$\vdash_0$" to denote the relation of intelim deducibility both for signed and unsigned formulae and write "$X \vdash_0 \varphi$", respectively "$\Gamma \vdash_0 A$", for "φ is 0-depth deducible from X", respectively "A is 0-depth deducible from Γ". We also write "$X \vdash_0$" ($\Gamma \vdash_0$) for "X is 0-depth refutable" ("Γ is 0-depth refutable").

Given a signed formula φ, let its unsigned version $\text{Uns}(\varphi)$ be defined as follows:

$$\text{Uns}(\varphi) = \begin{cases} A & \text{if } \varphi = T A \\ \neg A & \text{if } \varphi = F A, \end{cases}.$$

1	$T\,(p \vee q) \to \neg r$	Premise
2	$T\,p$	Premise
3	$T\,(p \wedge t) \to r$	Premise
4	$T\,p \vee q$	$T \vee \mathcal{I}_1, 2$
5	$T\,\neg r$	$T \to \mathcal{E}_1, 1, 4$
6	$F\,r$	$T\neg\mathcal{E}, 5$
7	$F\,(p \wedge t)$	$T \to \mathcal{E}_2, 3, 6$
8	$F\,t$	$F \wedge \mathcal{E}_1, 7, 2$

1	$(p \vee q) \to \neg r$	Premise
2	p	Premise
3	$p \wedge t \to r$	Premise
4	$p \vee q$	$T \vee \mathcal{I}_1, 2$
5	$\neg r$	$\to \mathcal{E}_1, 1, 4$
6	$\neg(p \wedge t)$	$\to \mathcal{E}_2, 3, 5$
7	$\neg t$	$\neg \wedge \mathcal{E}_1, 6, 2$

Figure 1.2: The sequence on the top is an intelim sequence using the rules for signed formulae. The one on the bottom is the corresponding sequence using the rules for unsigned formulae.

1.4. INTELIM SEQUENCES

We also write $\text{Uns}(X)$ for $\{\text{Uns}(\varphi) \mid \varphi \in X\}$. Then, it is not difficult to see that

$$X \vdash_0 \varphi \text{ if and only if } \text{Uns}(X) \vdash_0 \text{Uns}(\varphi).$$

The reader can verify that all the intelim rules are sound, namely that the conclusion is a 0-depth logical consequence of the premises. As for completeness, there is a technical point which must be taken care of. It follows from SCP and Definition 1.3.8 that in every shallow information state such that $v(A \vee A) = 1$, it must hold true that $v(A) = 1$. Similarly, in every shallow information state such that $v(A \wedge A) = 0$ it must hold true that $v(A) = 0$. Indeed, the value of A is by all means dictated, in both cases, by the value of $A \vee A$ ($A \wedge A$) and by the intended meaning of $\vee$ ($\wedge$) — as specified by the meaning constraints — via SCP. However, pure intelim sequences do not allow us to infer $T\, A$ from $T\, A \vee A$, or $F\, A$ from $F\, A \wedge A$, and so they are not *complete*. This technical problem can be addressed in three different ways:

1. We change the definition of well-formed formula of $\mathcal{L}$ as follows:

   - every atomic formula is a well-formed formula;
   - if A is a well-formed formula, so is $\neg A$;
   - if A and B are *distinct* well-formed formulae, so are $A \wedge B$ and $A \vee B$;
   - if A and B are (possibly identical) well-formed formulae, so is $A \to B$.

2. We stick to the standard definition of $\mathcal{L}$, pre-process all formulae and replace every occurrence of $A \vee A$ and $A \wedge A$, with A.

3. We introduce the two sound "mingle" rules in Table 1.8 in addition to the intelim rules; this solution is the same as the one adopted in (Finger & Gabbay, 2006) in response to a similar problem arising in their investigations into tractable subsystems of classical propositional logic.

All these solutions appear quite reasonable, although not terribly elegant. The first two amount to restricting the language without undermining its expressive power with respect to the constraint-based semantics. The third

$$\frac{TA \lor A}{TA} \; T\lor\text{-mingle} \qquad\qquad \frac{FA \land A}{FA} \; F\land\text{-mingle}$$

Table 1.8: Mingle rules for $\lor$ and $\land$.

one requires additional *ad hoc* rules that somehow spoil the harmony of the intelim approach, but are perfectly justified by the aim of extracting all the information that can possibly be obtained on the sole basis of the meaning of the logical operators. If we adopt any of them, intelim sequences turn out to be complete with respect to $\vDash_0$.

1.4.3 Another version of the single candidate principle

We shall now introduce a technical variant of the constraint-based semantics for which pure intelim sequences are sound and complete with no need for restricting the language or adding ad hoc rules, at the price of making the mingle rules unsound. In this variant, the SCP is not applied to valuation modules, intended as mappings from an $\mathcal{L}$-module to **3**, but directly to vectors of values in **3**. These can be thought of as rows of the truth-table of a Boolean operator in which some or all Boolean values may be replaced by the indeterminate value $\bot$.

Given a valuation module $\alpha : M \to \mathbf{3}$, such that its top formula is $A = \star(B_1, \ldots, B_n)$, the *value-vector of* α is the tuple:

$$\boldsymbol{\alpha} = \langle \alpha(B_1), \ldots, \alpha(B_n), \alpha(A) \rangle.$$

So, the i_{th} component of $\boldsymbol{\alpha}$, denoted by α_i, is equal to $\alpha(B_i)$ if $1 \leq i \leq n$, and to $\alpha(A)$ if $i = n + 1$. For example, Table 1.9 represents the value-vectors of all possible valuation modules α for the $\mathcal{L}$-module specified in the top row. Some of the rows are incompatible with the truth-table for $\land$, that is, with the graph of the Boolean function $f_\land : \{0,1\}^2 \to \{0,1\}$ that represents the classical meaning of $\land$. Such rows represent *inadmissible* vectors, while all the others are admissible. In particular, all the rows with 1 in the last column and 0 in some of the previous columns represent vectors that do not approximate any vector in $f_\land$, in the sense that replacing the undefined values with defined ones can never yield a vector in $f_\land$. To avoid unnecessary notational proliferation, we use for the approximation relation

1.4. INTELIM SEQUENCES

	p	q	$p \wedge q$
1	$\bot$	$\bot$	$\bot$
2	$\bot$	$\bot$	0
3	$\bot$	$\bot$	1
4	$\bot$	0	$\bot$
5	$\bot$	0	0
6	$\bot$	0	1
7	$\bot$	1	$\bot$
8	$\bot$	1	0
9	$\bot$	1	1
10	0	$\bot$	$\bot$
11	0	$\bot$	0
12	0	$\bot$	1
13	0	0	$\bot$
14	0	0	0
15	0	0	1
16	0	1	$\bot$
17	0	1	0
18	0	1	1
19	1	$\bot$	$\bot$
20	1	$\bot$	0
21	1	$\bot$	1
22	1	0	$\bot$
23	1	0	0
24	1	0	1
25	1	1	$\bot$
26	1	1	0
27	1	1	1

Table 1.9: Value-vectors for the $\mathcal{L}$-module $M = \{p, q, p \wedge q\}$.

between value-vectors the same symbol we used for the analogous relation between partial valuations. So, $\alpha \sqsubseteq \alpha'$ if α and α' have the same number of entries and $\alpha_i = \bot$ or $\alpha_i = \alpha'_i$.

Let $\dot{f}_\star$ be the subset of $f_\star$ containing all the vectors that approximate some vector in $f_\star$. In general, a valuation module α is inadmissible if its value vector $\alpha \notin \dot{f}_\star$, where $\star$ is the main operator of α. Observe that all inadmissible modules of Table 1.4 are inadmissible also under this new criterion. In Table 1.9 some admissible modules are such that replacing $\bot$ with one of the two defined values yields a vector that is inadmissible. For example, replacing the occurrence of $\bot$ with 1 in the eighth row, would yield an inadmissible vector. Hence, the piece of information $\langle p, 0 \rangle$ is implicitly contained in the value vector $\alpha = \langle 1, \bot, 0 \rangle$.

In this setting, the SCP takes the following form: given a valuation module α with main operator of arity n, a formula A such that $\alpha(A) = \alpha_i = \bot$ ($1 \leq i \leq n+1$), and ($x \in \{0,1\}$):

(SCP') $\qquad \alpha \Vdash_{\text{SCP}'} \langle A, x \rangle \iff \alpha[\alpha_i := 1-x] \notin \dot{f}_\star,$

where $\alpha[\alpha_i := 1-x]$ denotes the result of replacing the value $\alpha_i = \bot$ in α with $1-x$.

Again, a module is *stable* if and only if it is closed under SCP' and an admissible valuation is a *shallow information state* if all its modules are stable.

Observe that, in this new setting, the mingle rules (Table 1.8 above) are unsound, while the intelim rules are all sound. Indeed, given the valuation module α such that $\alpha(p \vee p) = 1$ and $\alpha(p) = \bot$, the value-vector $\langle \bot, 0, 1 \rangle$ is admissible and so is the value-vector $\langle 0, \bot, 1 \rangle$, for both approximate some vector in the graph of $f_\vee$. Similarly, given the valuation module α such that $\alpha(p \wedge p) = 0$ and $\alpha(p) = \bot$, the value-vector $\langle \bot, 1, 0 \rangle$ is admissible and so is the value-vector $\langle 1, \bot, 0 \rangle$.

1.4.4 Soundness and completeness

Let us rename "$\vDash_0^1$" the relation $\vDash_0$ of Definition 1.3.10 and let $\vDash_0^2$ be defined as $\vDash_0^1$ except that the notion of shallow information state is construed in terms of SCP' rather than SCP.

Let us call *intelim** the set consisting of the intelim rules in Table 1.5 and of the mingle rules in Table 1.8 and assume the notions of intelim* sequence,

1.4. INTELIM SEQUENCES

intelim* deduction and intelim* refutation are defined as in Definition 1.4.2 except for replacing "intelim" with "intelim* ".

Proposition 1.4.3. *For every finite set X of signed formulae and every signed formula φ,*

- $X \vDash_0^1 \varphi$ *($X \vDash_0^1$) if and only if there is an intelim* deduction of φ from X (an intelim* refutation of X).*

- $X \vDash_0^2 \varphi$ *($X \vDash_0^2$) if and only if there is an intelim deduction of φ from X (an intelim refutation of X).*

A proof is sketched in Section A.1 of the Appendix. The version for unsigned formulae is obtained in the usual way. We can use $\vdash_0^1$ and $\vdash_0^2$ to denote, respectively, intelim* and pure intelim deducibility to mirror the consequence relations $\vDash_0^1$ and $\vDash_0^2$. However, since most of the definitions and proofs are essentially the same in the two cases, in the sequel we shall simply use $\vDash_0$ to refer to both consequence relations and $\vdash_0$ to refer to both deducibility relations, except when otherwise specified. In what follows, we shall also use the expression "intelim" ambiguously for both "intelim" and "intelim*". Moreover, we shall use the expression "SCP" to refer to both variants, SCP and SCP'.

1.4.5 Subformula property and tractability

We extend Definition 1.3.2 to S-formulae in the obvious way:

Definition 1.4.4. *$S_1 A$ is a (proper, immediate) signed subformula of $S_2 B$ if, and only if, A is a (proper, immediate) subformula of B.*

The *subformula property* (SFP) is an all-important property of proofs and refutations that plays a crucial rôle in the automation of logical reasoning.

Definition 1.4.5. *An intelim proof of φ from X (an intelim refutation of X) has the SFP if every S-formula that occurs in it is a signed subformula of some S-formula in $X \cup \{\varphi\}$ (in X).*

While elimination rules preserve the subformula property — the conclusion of their application is always a signed subformula of the major premise

— introduction rules do not. So, it is important, both for theoretical and practical reasons, to show that the search for intelim proofs or refutations can be restricted, without any loss, to sequences involving only signed subformulae of the assumptions or, in the case of proofs, of the conclusion. As far as $\vdash_0$ is concerned, the only rules that may lead to violations of the SFP are the introduction rules. These rules prompt us for bounding the "search space", i.e., the set of the formulae that we need to consider in the search for a proof or a refutation. Moreover, an improper use of the introduction rules is related to the explosivity of $\vdash_0$:

Proposition 1.4.6. *If X is intelim inconsistent, then $X \vdash_0 \varphi$ for every signed formula φ.*

To see this, suppose X is intelim-inconsistent. Then there is a closed intelim sequence for X, i.e., $X \vdash_0 T A$ and $X \vdash_0 F A$ for some formula A. Now, if $\varphi = T B$ for some arbitrary formula B, then infer $T A \vee B$ from $T A$, by $T\vee\text{-}\mathcal{I}1$ (see Table 1.5), and $T B$ from $T A \vee B$ and $F A$ by $T\vee\text{-}\mathcal{E}1$.

If $\varphi = F B$ for some arbitrary B, infer $F A \wedge B$ from $F A$ using $F \wedge \text{-}\mathcal{I}$, and then $F B$ from $F A \wedge B$ and $T A$, using $F \wedge -\mathcal{E}1$.

Observe that these proofs do not have the SFP whenever $T A \vee B$ or $F A \wedge B$ are not signed subformulae of S-formulae in X, as is the case when, for example, $X = \{T A, F A\}$ and the arguments are represented by the following intelim sequences:

$$
(1.6) \quad
\begin{array}{ll}
T A & T A \\
F A & F A \\
T A \vee B & F A \wedge B \\
T B & F B.
\end{array}
$$

The intelim sequence on the left is nothing but a version for signed formulae of the argument that is often called "Lewis' proof" in the literature. The version for unsigned formulae is obtained by replacing each signed formula φ with its unsigned version $\text{Uns}(\varphi)$ and using the rules in Table 1.6.

We now show that the logic $\vdash_0$ does satisfy the subformula property in the following form:

1. if X is intelim inconsistent, there is an intelim refutation of X that contains only signed subformulae of S-formulae in X;

1.4. INTELIM SEQUENCES

2. if X is intelim consistent and φ is intelim deducible from X, there is an intelim proof of φ from X that contains only signed subformulae of S-formulae in $X \cup \{\varphi\}$.

This version of the subformula property is sufficient for all practical purposes. For, if we are trying to refute X we know that the search space can be restricted to intelim sequences that contain only signed subformulae of the input S-formulae: X is intelim inconsistent if and only if one of these sequences is closed. On the other hand, if we are trying to prove φ from X the search space can be restricted to intelim sequences that contain only signed subformulae of the input S-formulae in $X \cup \{\varphi\}$: $X \vdash_0 \varphi$ if and only if one of these sequences is closed or ends with φ. In the former case, we can take the closed sequence as a "proof" of any arbitrary S-formula φ, knowing that, by Proposition 1.4.6, it can always be extended to one that ends with φ. (On the other hand, one may wonder whether this makes any sense at all, and be content with saying that X is inconsistent.)

Observe that $\vdash_0$ restricted to intelim sequences with the SFP *is not* explosive, and so it is not complete with respect to $\vDash_0$ (see p. 34), albeit it always delivers a proof whenever the premises are 0-depth consistent. In order to achieve completeness, we can modify the definition of intelim proof as follows:

Definition 1.4.7. *An intelim sequence π for X is an* intelim proof *of φ from X if either (i) π is closed, or (ii) φ is the last S-formula in the sequence. In case (i) we say that the proof is* improper.

Clearly, improper "proofs" are such only in a Pickwickian sense. We could even call them *Pickwickian proofs*.

Consider now the following intelim sequences:

1	$T p \to \neg q$	Assumption	1	$T p$	Assumption
2	$T (p \to \neg q) \to p$	Assumption	2	$T \neg p$	Assumption
3	$T p \to r$	Assumption	3	$F p$	from 2
4	$T p$	Assumption	4	$T p \vee q$	from 1
5	$T \neg q$	from 1,4	5	$T q$	from 3, 4
6	$F q$	from 5			
7	$T p \vee q$	from 4			
8	$T p$	from 7,6			
9	$T r$	from 3,8			

The first one is an intelim proof of $T\,r$ from

$$\{T\,p \to \neg q, T\,(p \to \neg q) \to p, T\,p \to r\},$$

and the second one is again the usual proof of (an arbitrary) $T\,q$ from $\{T\,p, T\,\neg p\}$, which is often used to show the explosivity of classical logic. Observe that both proofs are *redundant*:

- In the first proof the S-formula $T\,p \vee q$ is first introduced (from premise $T\,p$) and then eliminated (using the minor premise $F\,q$) to re-obtain the S-formula $T\,p$, which was *already* contained in the sequence; that is, this proof contains circular reasoning;

- in the second proof, the S-formula $T\,p \vee q$ is first introduced (from premise $T\,p$) and then eliminated (using $F\,p$ as minor premise); however, the sequence was already closed before the $T\vee$-introduction and so, by Definition 1.4.7, the closed sequence $T\,p, T\,\neg p, F\,p$ was *already* a (Pickwickian) proof of $T\,q$ from $T\,p$ and $T\,\neg p$.

The same kind of redundancy is observed whenever a formula is, at the same time, the conclusion of an introduction and the major premise of an elimination.

Definition 1.4.8. *Given an intelim sequence π for X, a signed formula φ occurring in π is a* detour *if φ occurs in π both as conclusion of an introduction and as major premise of an elimination.*

Let us say that two sets X and Y of signed formulae *clash* if there is a formula A such that $SA \in X$ and $\overline{S}A \in Y$. The following lemma is key to the SFP of intelim proofs and refutations. It implies that detours can be safely removed from intelim proofs and refutations.

Lemma 1.4.9 (Inversion Principle). *If a conclusion φ follows from $X \cup \{\psi\}$ by an application of an elimination rule, with ψ as major premise, then φ belongs to every Y such that (i) ψ follows from Y by an application of an introduction rule, and (ii) Y does not clash with X.*

For the standard Boolean operators the above lemma can be easily established by inspection of the intelim rules. For the general case of an arbitrary Boolean operator see (D'Agostino et al., 2013).

1.4. INTELIM SEQUENCES

In other words: by an elimination we do not obtain any (consistent) information that would not be already available if its major premise had been obtained by an introduction. An inversion principle similar to the one expressed in Lemma 1.4.9 is well-known in the context of Gentzen's and Prawitz's Natural Deduction and is taken as a "justification" of the elimination rules in terms of the introduction rules. It is interesting that a principle of the same kind holds also in our setting, despite its being considerably different from Gentzen's and Prawitz's.

Definition 1.4.10. *An intelim sequence π is* non-redundant *if both the following conditions are satisfied:*

1. *π contains no more than one occurrence of the same S-formula,*

2. *π does not properly contain a closed intelim sequence.*

Otherwise, we say that it is redundant.

Lemma 1.4.11. *If π is non-redundant, π contains no detours;*

Proof. Suppose π contains a detour, namely a formula φ that is at the same time the conclusion of an introduction and the major premise of an elimination. By Lemma 1.4.9, either the conclusion of the elimination is equal to one of the premises of the introduction, or one of the minor premises of the elimination is the conjugate of one of the premises of the introduction, and so the sequence was already closed before the elimination. In either case π is redundant. □

Observe that:

Fact 1.4.12. *If π is a non-redundant intelim sequence containing a pair of conjugate S-formulae, it is not the case that both of them result from the application of an introduction rule.*

An example will suffice. Suppose that both $T\,A \wedge B$ and $F\,A \wedge B$ are conclusions of introductions in π. Then, both $T\,A$ and $T\,B$ and at least one of $F\,A$ and $F\,B$ already belong to π. Then, π was already closed before the introductions and therefore is redundant.

Definition 1.4.13. *An occurrence of an S-formula φ in an intelim sequence π for X is* idle *if (i) $\varphi \notin X$ and is not the last S-formula of π, (ii) it is not used in π as premise of some application of an intelim rule, and (iii) it is not the conjugate of the last S-formula in the sequence.*

Definition 1.4.14. *An intelim sequence π is* strongly non-redundant *if it is non-redundant and contains no idle occurrences of S-formulae.*

Proposition 1.4.15. *If π is a strongly non-redundant intelim proof of φ from X (refutation of X), then π has the SFP.*

A proof is in Section A.2 in the Appendix.

Remark 1.4.16. *It is easy to see that every intelim sequence can be turned into a* shorter *strongly non-redundant one.*

Let us define the *size* of a sequence π of signed formulae, denoted by $|\pi|$, as the total number of occurrences of symbols in π.

Corollary 1.4.17 (Subformula Property). *For all X and φ,*

1. *If there is an intelim refutation π of X, then there is an intelim refutation π' of X with the SFP such that $|\pi'| \leq |\pi|$;*

2. *if there is an intelim proof π of φ from X, then there is an intelim sequence π' such that $|\pi'| \leq |\pi|$ that is either an intelim proof of φ from X with the SFP or a refutation of X (i.e., an improper proof of φ from X) with the SFP.*

The corollary follows immediately from Proposition 1.4.15 and from Remark 1.4.16.

As the corollary suggests, there are X and φ such that φ is intelim deducible from X by means of an intelim proof, but there is *no* intelim proof of φ from X with the SFP. This may be the case if the intelim sequence contains detours, and so is redundant (by Lemma 1.4.11), as the second proof on p. 49. Such proofs can, however, be turned into refutations of the premises, that is into improper ("Pickwickian") proofs of the conclusion, or of any conclusion for that matter, with the SFP.

In the proof-theoretical literature, proofs and refutations with the SFP are often said to be "analytic". This syntactic sense of "analytic" goes back

1.4. INTELIM SEQUENCES

to (Gentzen, 1969) and is not to be confused with its semantic or informational sense (see Chapter 4).[19]

Intelim proofs and refutations for unsigned formulae enjoy the SFP in a weaker form.

Definition 1.4.18. *A formula A is a* weak subformula *of a formula B if it is either a subformula of B or the negation of a subformula of B.*

Lemma 1.4.9 can be adapted to the intelim rules for unsigned formulae. In this case, non-redundant proofs and refutations enjoy the *weak subformula property* (WSFP): every formula occurring in them is a weak subformula of the input formulae.

It is interesting to observe that, by Corollary 1.4.17, syntactically analytic intelim proofs or refutations are *uniformly shorter* than non-analytic ones. This property can be contrasted with the well-known fact that syntactically analytic proofs or refutations in full classical propositional logic may be exponentially longer than non-analytic ones.

Observe that Corollary 1.4.17 and Proposition 1.4.3, taken together, imply that intelim deducibility and refutability can be characterized in terms of shallow information states whose domain is not the whole set of all $\mathcal{L}$-formulae, but is restricted to the relevant subformulae. Let $\operatorname{sub}(\Gamma)$ denote the set of all subformulae of the formulae in Γ.

Definition 1.4.19. *A set Δ of $\mathcal{L}$-formulae is an $\mathcal{L}$-*domain *if $\Delta = \operatorname{sub}(\Delta)$.*

We can then consider *syntactically bounded partial valuations*, namely mappings $v : \Delta \to \mathbf{3}$ where Δ is an $\mathcal{L}$-domain and extend all the semantic definitions in the obvious way. In particular:

---

[19] Syntactically analytic proofs, that is proofs with the SFP, are often incorrectly identified with cut-free proofs in Gentzen's sequent calculus. Clearly, the latter do enjoy the SFP, but proofs with the SFP need not be cut-free, since cut can be restricted to subformulae. In the sequel we shall show that adding a rule corresponding to the classical cut rule and allowing its unbounded use makes the intelim system complete for classical propositional logic and the analog of the cut rule is not redundant, but can be restricted to subformulae. In this way (analytic) cut is applied only when the intelim rules cannot be further applied. Such intelim proofs with analytic cut are exponentially more efficient than cut-free proofs in Gentzen's sequent calculus (D'Agostino, 1992; D'Agostino & Mondadori, 1994).

Definition 1.4.20. *For every $\mathcal{L}$-domain Δ, a* shallow information state *over Δ is a mapping $v : \Delta \to \mathbf{3}$ such that every module of v is stable (closed under SCP).*

Corollary 1.4.21. *Given a set Γ of formulae and a formula A,*

- *$\Gamma \vdash_0 A$ if and only if $\Gamma \models_0 A$ if and only if $v(A) = 1$ for all shallow information states v over $\mathtt{sub}(\Gamma \cup \{A\})$ such that $v(B) = 1$ for all $B \in \Gamma$;*

- *$\Gamma \vdash_0$ if and only if $\Gamma \models_0$ if and only if there is no shallow information state over $\mathtt{sub}(\Gamma)$ such that $v(B) = 1$ for all $B \in \Gamma$.*

We have chosen to express the above corollary in terms of unsigned formulae, but it is routine to express it in terms of S-formulae. This restriction of the search space makes exploring it for potential applications of intelim rules a task that is computationally (and cognitively) easy. Let $|A|$ denote the size of the formula A, i.e. the total number of occurrences of symbols in it, and $|\Delta|$ the total number of occurrences of symbols in the formulae of Δ.

Proposition 1.4.22. *Whether or not A is intelim deducible from Γ (Γ is intelim refutable) can be decided in time $O(n^2)$, where $n = |\Gamma \cup \{A\}|$ ($n = |\Gamma|$).*

A simple algorithm can be found in (D'Agostino et al., 2013) and is reported in Section A.3 of the Appendix. Here we illustrate the procedure with an example.

Example 1.4.23. *We want to decide whether $\neg t$ is intelim deducible from the following set of assumptions:*

(1.7) $\qquad \Gamma = \{p \vee q, p \to (r \wedge s), q \to \neg(t \wedge \neg s), \neg s\}$

We start by constructing the subformula graph *for the input set, that is, the graph whose nodes are the formulae in $\mathtt{sub}(\Gamma \cup \{\neg t\})$ and whose edges connect two formulae whenever one is an immediate subformula of the other. Next we label all the formulae in the graph with their current informational value. So, we start by labelling all the premises in Γ with "1" and all the

1.4. INTELIM SEQUENCES 55

other formulae with "?" to express the fact that their informational value is undefined (in this context we find the question mark more suggestive than "⊥"). This is the initial graph that represents the information that we explicitly hold.

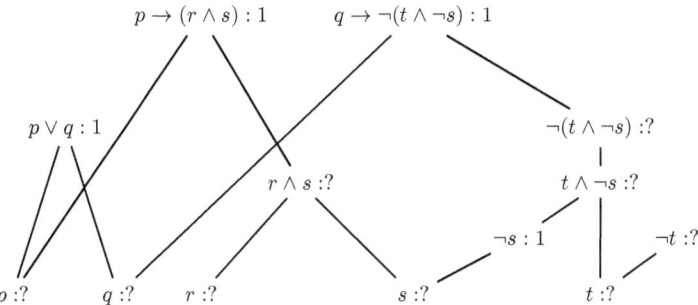

Given an arbitrary order of the defined formulae (we assume the order in which they are listed in (1.7)) we start by visiting the first defined formula, namely $p \vee q$ and examine the valuation modules to which it belongs. In this case the only one is the module of which $p \vee q$ is the top formula. Since the module is stable (the SCP cannot be applied), we proceed to the next defined formula ($p \rightarrow (r \wedge s)$). Again, the only module to which it belongs is stable. The same applies to $q \rightarrow \neg(t \wedge \neg s)$. The following defined formula is $\neg s$, but now one of the two valuation modules to which it belongs is unstable on s. So, by the SCP (or, equivalently by the associated elimination rule $T \neg \text{-} \mathcal{E}$) we update the label of s, changing its informational value from "?" to "0".

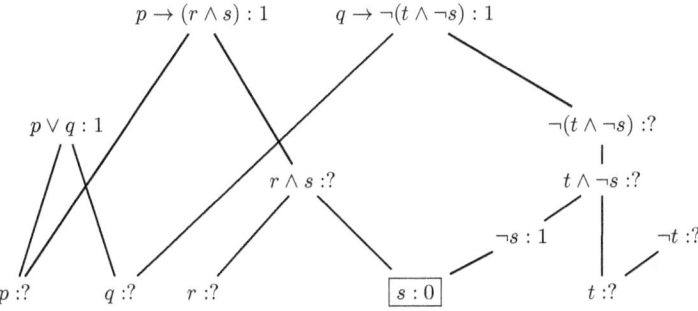

Next we visit the newly defined formula s (in the framed node) and verify

that one of the modules to which it belongs is unstable, namely the one with top formula $r \wedge s$. By the SCP (or, equivalently, by the introduction rule $F \wedge \text{-} \mathcal{I}2$) we update the label of $r \wedge s$ so as to make the module stable.

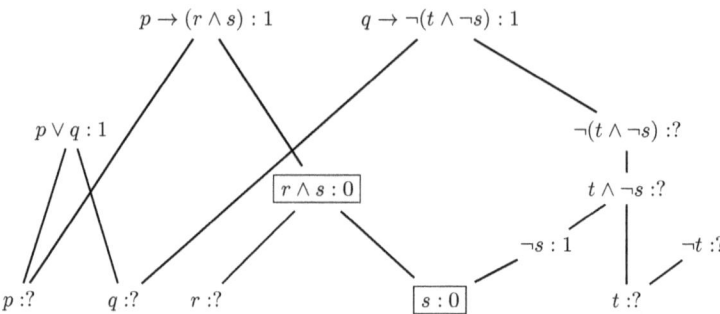

We proceed as before, visiting the newly defined formulae and checking only the valuation modules to which they belong and applying the SCP (or equivalently the associated intelim rules) to the unstable modules. The result is the following sequence of updates:

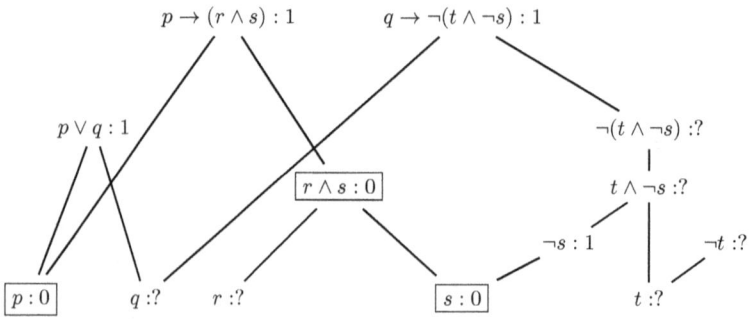

1.4. INTELIM SEQUENCES

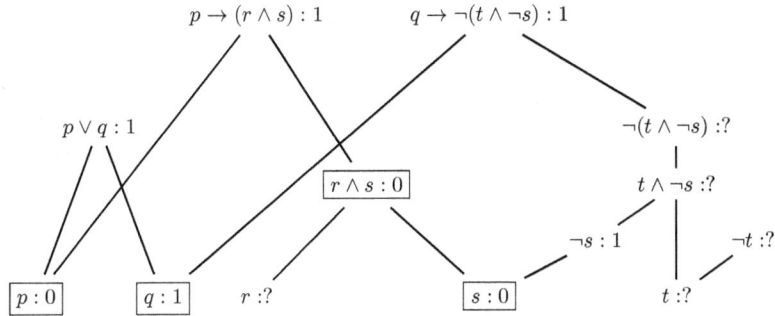

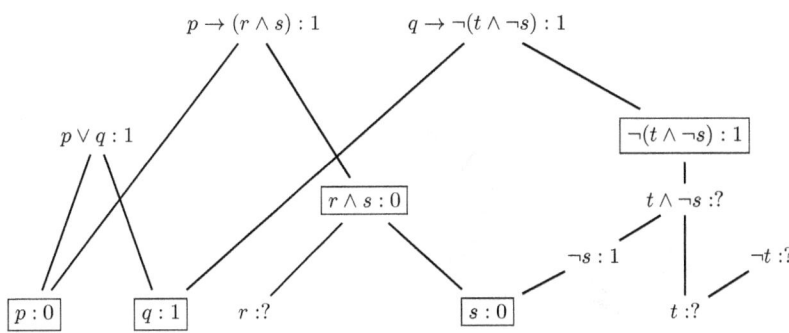

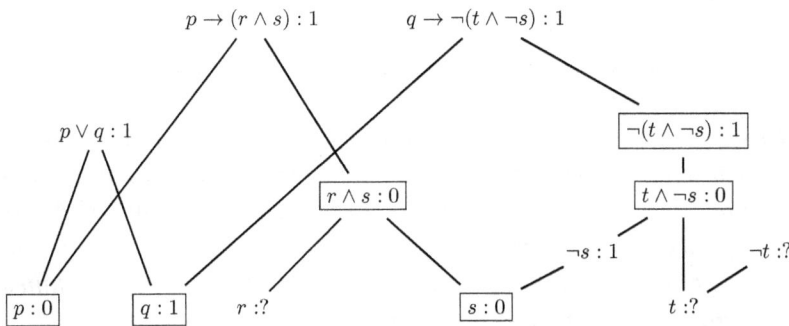

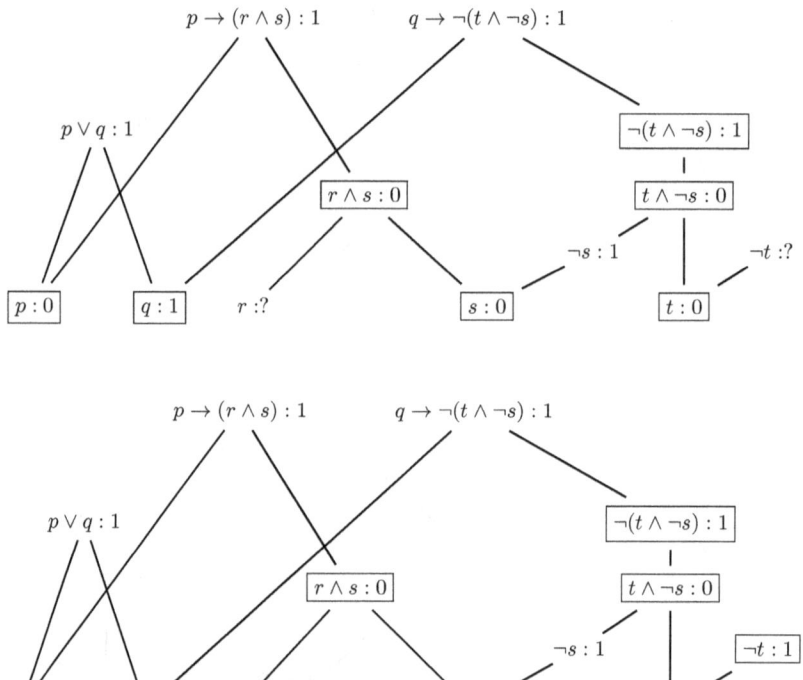

The last labelled graph represents a valuation of sub($\Gamma \cup \{\neg t\}$) *in which all valuation modules are stable. The framed nodes represent the implicit information content of the premises with respect to the $\mathcal{L}$-domain specified by the graph, i.e.,* sub(Γ) $\cup \{\neg t\}$).

This final graph is saturated with respect to applications of the SCP (aka, the intelim rules) and can be immediately transformed into a normal intelim proof of $T \neg t$ from $T (\Gamma)$ as well as into a normal proof of $\neg p$ from Γ, using unsigned formulae, in the usual way. In this example the proof would be strongly non-redundant; in the general case, the graph may contain some idle formulae (but in any case no detours).

A crucial observation is that every defined formula needs to be visited at most once in oder to make the graph stable. Suppose a defined formula does not trigger an update, because every valuation module in which it is contained is stable, and as a result of a subsequent update one of these modules becomes unstable, then this instability will be resolved when visiting the

1.5. NON-DETERMINISTIC SEMANTICS

newly defined formula that has caused it. For example, consider the module with top formula $p \vee q$ that is initially stable. When it becomes unstable, i.,e. when the label of p is updated to 0, the instability is solved anyway when visiting the newly defined p. So, there is no need to visit the same node more than once, provided that we visit only nodes containing defined formulae. Moreover, the order in which the defined nodes are visited does not make any difference and the final stable graph is unique.

1.5 Non-deterministic semantics

1.5.1 Valuation systems

Given a propositional language $\mathcal{L}$ with logical operators $\star_1, \ldots, \star_n$, a *valuation system* (or *matrix*) for $\mathcal{L}$ is a structure $\mathcal{V} = \langle V, D, f_1, \ldots, f_n \rangle$, where:

1. V is a set with at least two elements;

2. D is a proper non-empty subset of V;

3. for each i, $1 \leq i \leq n$, f_i is a mapping of V^{a_i} to V, where a_i is the arity of $\star_i$.

A valuation system is *finite*, if V contains a finite number of elements. A *Boolean valuation system* is one in which $V = \{0, 1\}$ and $D = \{1\}$. An *assignment* relative to a valuation system $\mathcal{V}$ for $\mathcal{L}$ is a mapping of the atomic formulae of $\mathcal{L}$ to V. Each assignment ϕ induces the (total) *valuation* v_ϕ over V defined as follows:

- $v_\phi(p) = \phi(p)$ for all atomic p,

- $v_\phi(\star_i(A_1, \ldots, A_{a_i})) = f_i(v_\phi(A_1), \ldots, v_\phi(A_{a_i}))$.

For every set Γ of $\mathcal{L}$-formulas and every $\mathcal{L}$-formula A, $\Gamma \models_\mathcal{V} A$ (Γ *entails* A in $\mathcal{V}$) if and only if, for every assignment ϕ, $v_\phi(A) \in D$ whenever $v_\phi(B) \in D$ for all $B \in \Gamma$. Given a propositional logic $L = \langle \mathcal{L}, \models_L \rangle$ and a valuation system $\mathcal{V}$ for $\mathcal{L}$, we say that

- $\mathcal{V}$ is *faithful* to L if, and only if, for every Γ and A, $\Gamma \models_L A$ implies $\Gamma \models_\mathcal{V} A$;

- $\mathcal{V}$ is *characteristic* for L if, and only if, $\mathcal{V}$ is faithful to L and, for all Γ and A, $\Gamma \models_{\mathcal{V}} A$ implies $\Gamma \models_L A$.

1.5.2 Lack of a finite valuation system for the 0-depth logic

We now show that the modular semantics described in Sections 1.3.3 and 1.4.3 is not equivalent to any standard finitely-valued semantics.

Proposition 1.5.1 (D'Agostino *et al.*, 2013). *There is no finite valuation system characteristic for* $\langle \mathcal{L}, \models_0 \rangle$.

Proof. Let $\mathcal{V}$ be a valuation system such that V contains n elements. Given $n+2$ distinct atomic formulas $p_0, \ldots, p_{n+1}$ of $\mathcal{L}$, let $\Gamma = \{p_i \to p_{i+1}\}_{0 \le i \le n}$ and let

$$C = \bigvee_{2 \le j \le n+1} \bigvee_{0 \le i \le j-2} p_i \to p_j.$$

For example, when $n = 3$,

$$\Gamma = \{p_0 \to p_1, p_1 \to p_2, p_2 \to p_3, p_3 \to p_4\}$$

and

$$C = (p_0 \to p_2) \vee (p_0 \to p_3) \vee (p_1 \to p_3) \vee (p_0 \to p_4) \vee (p_1 \to p_4) \vee (p_2 \to p_4).$$

Now, let ϕ be an assignments such that $v_\phi(B) \in D$ for all $B \in \Gamma$. Observe that, since there are only n distinct elements in V, then for some j, $2 \le j \le n+1$, and some i, $0 \le i \le j-2$, it must be the case that $\phi(p_j) = \phi(p_{i+1})$ and, therefore, $v_\phi(p_i \to p_j) = v_\phi(p_i \to p_{i+1}) \in D$. In the above example for $n = 3$, it is impossible that

$$\phi(p_2) \ne \phi(p_1)$$
$$\phi(p_3) \ne \phi(p_2) \qquad \phi(p_3) \ne \phi(p_1)$$
$$\phi(p_4) \ne \phi(p_3) \qquad \phi(p_4) \ne \phi(p_2) \qquad \phi(p_4) \ne \phi(p_1)$$

are all simultaneously true inequalities and, therefore, one of the disjuncts in C must receive the same designated value as one of the formulas in Γ.

1.5. NON-DETERMINISTIC SEMANTICS

Hence, since $\mathcal{V}$ is characteristic for $\langle \mathcal{L}, \models_0 \rangle$ and $A \models_0 A \vee B$ for all $A, B \in \mathcal{L}$, it follows that $v_\phi(C) \in D$ and so $\Gamma \models_\mathcal{V} C$. However, it is not difficult to check that, by the decision procedure for intelim deducibility, $\Gamma \nvdash_0 C$ and therefore, by the completeness of $\vdash_0$, $\Gamma \not\models_0 C$.

□

1.5.3 Non-deterministic tables

The lack of a finite valuation system for the 0-depth logic, does not prevent the latter from being tractable (Proposition 1.4.22). Indeed, it turns out that this tractable logic admits of a non-standard valuation system in which the evaluation of formulae on the basis of a given assignment is *non-deterministic*.

Let V be a set of truth-values. A *non-deterministic valuation* over V is a mapping $\hat{v} : F(\mathcal{L}) \to V$ such that (i) for every atomic $p \in F(\mathcal{L})$, $\hat{v}(p) \in V$ and (ii) for every n-ary operator $\star$ and every formula $A \in F(\mathcal{L})$ with $\star$ as main operator, $\hat{v}(\star(B_1, \ldots, B_n)) \in \hat{f}_\star(\hat{v}(B_1), \ldots, \hat{v}(B_n))$ where $\hat{f}_\star$ is some function $V^n \to 2^V \setminus \emptyset$.

Notice that, unlike standard valuations, non-deterministic valuations are not uniquely determined by an assignment ϕ to the atomic formulas of values in **3**. In general, for every such assignment ϕ we have a set of non-deterministic valuations $\hat{v}$ such that, for all atomic p, $\hat{v}(p) = \phi(p)$. On the other hand, every non-deterministic valuation is clearly a Boolean valuation whenever all the atomic formulae are assigned values in $\{0, 1\}$.

Non-deterministic valuations have been intensively studied by Arnon Avron (see Avron (1991); Avron & Lev (2001); Avron (2005a); Avron & Zamansky (2006); Avron (2005b) among others) and used as a tool for investigating proof-theoretical properties of Gentzen-style sequent calculi. Crawford & Etherington (1998) introduced 3-valued non-deterministic valuations as a tool for investigating tractable inference and claimed (without proof) that they provide an adequate semantics for (an extension of) unit resolution. We generalize Crawford and Etherington's semantics and show that this generalization can be justified in terms of the modular semantics of Section 1.4.3, so building a bridge between the two approaches.

For every n-ary operator $\star$, let $\hat{f}_\star$ be the function from $\mathbf{3}^n$ to $2^\mathbf{3} \setminus \emptyset$

defined as follows (recall that $\mathbf{3} = \{0, 1, \bot\}$):

(1.8) $y \in \hat{f}_\star(x_1, \ldots, x_n)$ iff $\langle x_1, \ldots, x_n, y \rangle$ is a stable element of $\mathcal{A}_\star$.

Examples 1.5.2. The following tables describe the functions $\hat{f}_\wedge, \hat{f}_\vee, \hat{f}_\rightarrow$ and $\hat{f}_\neg$ for the usual Boolean operators:

$\wedge$	1	0	$\bot$		$\vee$	1	0	$\bot$		$\rightarrow$	1	0	$\bot$		$\neg$	
1	1	0	$\bot$		1	1	1	1		1	0	0	$\bot$		1	0
0	0	0	0		0	1	0	$\bot$		0	1	1	1		0	1
$\bot$	$\bot$	0	$\bot,0$		$\bot$	1	$\bot$	$\bot,1$		$\bot$	1	$\bot$	$\bot,1$		$\bot$	$\bot$.

These are the tables introduced by Crawford & Etherington (1998) whose connection with the modular semantics of the previous section is given by (1.8). Essentially the same tables were proposed by Quine (1974) to define what he called "the primitive meaning of the logical operators" (see D'Agostino, 2014a).

The following tables describe the non-deterministic functions for *exclusive or* ($\oplus$), *Peirce's arrow* ($\downarrow$), *Sheffer's stroke* ($|$) and *equivalence* ($\equiv$), built in accordance with (1.8):

$\oplus$	1	0	$\bot$		$\downarrow$	1	0	$\bot$
1	0	1	$\bot$		1	0	0	0
0	1	0	$\bot$		0	0	1	$\bot$
$\bot$	$\bot$	$\bot$	$\bot,1,0$		$\bot$	0	$\bot$	$\bot,0$

$|$	1	0	$\bot$		$\equiv$	1	0	$\bot$
1	0	1	$\bot$		1	1	0	$\bot$
0	1	1	1		0	0	1	$\bot$
$\bot$	$\bot$	1	$\bot,1$		$\bot$	$\bot$	$\bot$	$\bot,1,0$.

Definition 1.5.3. *A* **3ND**-*valuation is a non-deterministic valuation* $\hat{v}$ *over* **3** *satisfying the following condition, for every n-ary operator* $\star$:

$$\hat{v}(\star(A_1, \ldots, A_n)) \in \hat{f}_\star(\hat{v}(A_1), \ldots, \hat{v}(A_n)).$$

It immediately follows from the above definitions that:

1.5. NON-DETERMINISTIC SEMANTICS

Proposition 1.5.4. *A partial valuation v for $\mathcal{L}$ is a shallow information state if and only if v is a* **3ND***-valuation.*

Therefore:[20]

Corollary 1.5.5. *For every Γ, A,*

1. *$\Gamma \models_0 A$ if, and only if, A is verified by all* **3ND**-*valuations that verify all the formulas in Γ;*

2. *$\Gamma \models_0$ (Γ is 0-depth inconsistent) if and only if there is no* **3ND**-*valuation that verifies all the formulas in Γ.*

1.5.4 Expressive completeness

Given a logic $L = \langle \mathcal{L}, \models_L \rangle$, we say that an n-ary logical operator $\circ$ of $\mathcal{L}$ is *definable* in terms of other logical operators $\star_1, \ldots, \star_k$ of $\mathcal{L}$ if for every formula A containing $\circ$ there exists an equivalent formula B built up from the atomic formulas occurring in A by the only means of the logical operators $\star_1, \ldots, \star_k$. But, what do we mean here by "equivalent"? Compare the following answers:

(1) if the logic admits of a standard valuation system, in the sense specified in Section 1.5.1, A is equivalent to B if, for every assignment ϕ, $v_\phi(A) = v_\phi(B)$;

(2) A and B are equivalent whenever $A \models_L B$ and $B \models_L A$, which amounts to saying that A and B have the same truth-table.

Sense (2) is weaker than sense (1), for when there are more than two values it may be the case that $v_\phi(A) \in D$ if and only if $v_\phi(B) \in D$, but $v_\phi(A) \neq v_\phi(B)$ for some ϕ.

Under the informational semantics, which admits of no standard valuation system, sense (1) does not apply, because valuations are not uniquely determined by assignments to the atomic formulae. Under the consequence relation $\models_0$, no finite set of operators is sufficient to define, even in the weaker sense (2), all the others. For example, $A \vee B$ is not definable, under $\models_0$, as $\neg(\neg A \wedge \neg B)$, for $A \vee B \nvDash_0 \neg(\neg A \wedge \neg B)$ and $\neg(\neg A \wedge \neg B) \nvDash_0 A \vee B$,

---

[20] Here we use "$\models_0$" in the sense of $\models_0^2$ of Section 1.4.4.

despite the fact that $\neg(A \vee B) \vdash_0 \neg A \wedge \neg B$ and $\neg A \wedge \neg B \vdash_0 \neg(A \vee B)$. Some sort of expressive completeness can, however, be retrieved by introducing a third sense, similar to sense (1), in which two formulae can be equivalent

(3) A and B are equivalent in a sense similar to sense (1) if and only if they have the same **3ND** table (see Section 1.5.3).

This, in turn, is always the case if and only if A and B have the same classical truth-table. However, formulae that are equivalent in this sense may not be equivalent in sense (2). For example $A \vee B$ and $\neg(\neg A \wedge \neg B)$ are equivalent in this sense, but not in sense (2).

This situation implies that, in general, the 0-depth consequence relation defines different logics for different choices of the logical operators in the language $\mathcal{L}$, since there is no subset of the logical operators that is expressively complete in sense (2).

Chapter 2

Depth-bounded deduction

2.1 Weak depth-bounded approximations

The content of Chapter 1 can be summarized by saying that $\models_0$ is the smallest Tarskian consequence relation that is closed under the intelim rules for the logical operators of $\mathcal{L}$. In this chapter we use $\models_0$ as a basis for defining several approximation systems for Boolean Logic, i.e., directed sets of consequence relations that indefinitely approximate it. Depending on the type of approximation, the successors of $\models_0$ may or may not be Tarskian consequence relations. Whatever the type, however, the operational rules are the same for all layers, namely those shown in Tables 1.5 and 1.6. Each subsequent layer is obtained by controlling the use of virtual information and introducing more complex structures than simple intelim sequences.

2.1.1 Virtual information

Shallow information states are valuations in which each module is closed under SCP in one of the two senses explained in Sections 1.3.3 and 1.4.3. This means, in essence, that, each valuation module α is closed under all the information that is *uniquely determined* by the Boolean truth table for the main operator of α and this is the most basic mechanism for extracting implicit information from the data represented by a partial valuation. Any "deeper" processing of the data must introduce and use information that is not uniquely determined by it. For example, to establish the following typical instance of "reasoning by cases"

(2.1)
$$\frac{p \vee q \quad p \to r \quad q \to r}{r}$$

we start with the partial valuation v such that

(2.2)
$$v(A) = \begin{cases} 1 & \text{if } A \in \{p \vee q, p \to r, q \to r\} \\ \bot & \text{otherwise.} \end{cases}$$

The only valuation modules of v that are significant for the deduction problem under consideration are the following:

p	q	$p \vee q$		p	r	$p \to r$		q	r	$q \to r$
$\bot$	$\bot$	1		$\bot$	$\bot$	1		$\bot$	$\bot$	1.

These valuation modules are all closed under SCP ("stable"): there is no way of determining the value of undefined formulae from the information explicitly stored in the module. In this sense all the modules are *stable* or *informationally closed*. So, v is a shallow information state. To extract the implicit information that r is true, we necessarily have to consider possible refinements of v containing information that is not even implicitly contained in it, i.e. what we have called *virtual information*. For example, we may consider its possible alternative refinements v_1 and v_2 such that $v_1(p) = 1$, $v_2(p) = 0$ and $v_1(A) = v_2(A) = v(A)$ for all $A \neq p$. Both such refinements of v contain information concerning p that is not contained in v. Depending on the context, we may assume that either v_1 or v_2 must "eventually" obtain, that is, one of the two valuations will eventually represent the information explicitly held by the agent at a certain point in the future. Or, alternatively, argue that either v_1 or v_2 (or both) must agree with the complete information state of an omniscient agent whenever v does. Else, we may think of $v_1(p)$ and $v_2(p)$ as the possible answers obtained by querying a reliable source about the value of p. In any case, the relevant modules of v_1 and v_2 are, respectively:

p	q	$p \vee q$		p	r	$p \to r$		q	r	$q \to r$
1	$\bot$	1		1	$\bot$	1		$\bot$	$\bot$	1

2.1. WEAK DEPTH-BOUNDED APPROXIMATIONS

and

p	q	$p \vee q$
0	$\bot$	1

p	r	$p \to r$
0	$\bot$	1

q	r	$q \to r$
$\bot$	$\bot$	1.

In both cases the framed module has become unstable and that the least valuation (with respect to the $\sqsubseteq$ relation) in which all the relevant modules are stable verifies r. So, we are allowed to conclude that the piece of information that r is true is also implicitly contained in v, albeit this step is more "costly" than a simple application of the SCP, in that it essentially requires the consideration of refinements of v that contain virtual information.

To take another example involving hypothetical reasoning, consider the following simple inference:

(2.3)
$$\frac{p \to q \quad q \to r}{p \to r.}$$

Again, we cannot obtain the conclusion from the premises without introducing virtual information. Consider a valuation v such that

(2.4)
$$v(A) = \begin{cases} 1 & \text{if } A \in \{p \to q, q \to r\} \\ \bot & \text{otherwise.} \end{cases}$$

The valuation modules involved are:

p	q	$p \to q$
$\bot$	$\bot$	1

q	r	$q \to r$
$\bot$	$\bot$	1

p	r	$p \to r$
$\bot$	$\bot$	$\bot$.

These valuation modules are all closed under SCP (stable) and so v is a shallow information state. Again, in order to draw the desired conclusion we have to consider possible refinements of v containing virtual information. Let us consider its possible alternative refinements v_1 and v_2 such that $v_1(p) = 1$, $v_2(p) = 0$ and $v_1(A) = v_2(A) = v(A)$ for all $A \neq p$. In this case, the relevant modules of v_1 and v_2 are, respectively,

p	q	$p \to q$
1	$\bot$	1

q	r	$q \to r$
$\bot$	$\bot$	1

p	r	$p \to r$
1	$\bot$	$\bot$

and

p	q	$p \to q$	q	r	$q \to r$	p	r	$p \to r$
0	$\bot$	1	$\bot$	$\bot$	1	0	$\bot$	$\bot$.

In both cases, as before, the framed module has become unstable and the least valuation (with respect to the $\sqsubseteq$ relation) in which all the relevant modules are stable verifies $p \to r$.

The need to introduce virtual information that is not contained in the premises in order to reach a conclusion has a certain Kantian flavour and appears to qualify this kind of argument as "synthetic" in some sense related to Kant's: it forces us to go *beyond the data*, to consider information that is not "given" in the premises and cannot be analytically extracted from them. Yet, the argument is *a priori* in that it does not depend on experience (on this point see Chapter 4). For this reason the conclusion cannot be obtained by means of a mere analysis of the meaning of the logical words, if we take this meaning in its most primitive version given by the negative constraints.

Remark 2.1.1. *Once inferences such as the ones in (2.1) and (2.3) have been established, the agent may well* learn *the pattern behind them and "lift" them to the status of "easy" 0-depth inference rules that can be added to the intelim rules. This will increase the deductive power of the basic layer without affecting tractability. Indeed, depending on the applications, adding a rule-learning component to the approximation system, may lead to decrease the average complexity of the depth-bounded approximations considerably.*

Apart from the most simple instances of case and hypothetical reasoning, another typical pattern that can be "learned" at depth 1 consists of *reductio ad absurdum*. It can be easily checked that there is no shallow information state in which the premises of (2.1) or of (2.3) are verified and the conclusion is falsified. Hence, the latter must be verified in any refinement of v in which the conclusion is defined.[1]

---

[1]This may look like an application of the SCP, and it may be observed that even the SCP makes some minimal use of virtual information. However, the SCP is applied locally, within a module, and allows us to extract the most superficial implicit information locally contained in the data, which can be reasonably regarded as "actual", in the sense of being (very) easily accessible.

2.1. WEAK DEPTH-BOUNDED APPROXIMATIONS

On the other hand, the principle used in the *reductio* for (2.1) and (2.3) is a sort of "higher degree" SCP, which is *not applied locally* within a module. So, the use of virtual information in these arguments makes them more complex and increases the "difficulty" of the task.

Remark 2.1.2. *While the 0-depth layer does not allow us to prove any tautology (see above, p. 35), each layer of depth $k > 0$ of the approximations we are going to define in this chapter allows us to prove increasingly complex tautologies. Interestingly enough, it appears that all the axioms of typical Hilbert style systems can be proven by reductio ad absurdum at depth 1, by showing that their negation is 0-depth inconsistent (so far we have been unable to find a counterexample). In other words they are among the easiest tautologies that can be recognized by virtue of the informational meaning of the logical operators, which appears to be in in tune with the intuitive idea of what an axiom should be*[2].

The unbounded use of virtual information, in the way just explained, turns an information state into a Boolean valuation. A natural way of approximating Boolean valuations, starting from the shallow information states defined in Chapter 1 consists in imposing an upper bound on the depth at which its nested use is allowed.

2.1.2 Virtual space and search space

A fine-grained control on the manipulation of virtual information involves a good deal of subtleties, both from the proof-theoretical and the semantical viewpoint. In this section we propose a general approach to deal with a variety of approximation strategies that differ from each other on two respects: (i) the delimitation of the *virtual space*, namely the set of the formulae that we are allowed to use to introduce virtual information and (ii) the delimitation of the *search space*, namely the set of all formulae we are "interested in" as potential premises and conclusions of our inference steps.

For each deduction or refutation problem the virtual space and the search space are defined by functions h of the set Δ of the formulae that occur in the problem specification, namely X and φ for problems of the form "$X \vDash_0 \varphi$?"

---
[2] We owe this observation to David Makinson and Lloyd Humberstone, personal communication.

and just X for problems of the form "$X \models_0$?". In the sequel we shall use the expressions "input formulae" or "input set" to refer to these formulae. The search space and the virtual space must be closed under subformulae (i.e., they must be $\mathcal{L}$-domains), that is:

(2.5) $$\mathtt{sub}(h(\Delta)) = h(\Delta),$$

where sub is the function that, when applied to Δ, yields the set of all the subformulae of the formulae contained in Δ. Moreover, the search space must obviously include the input set, that is, for every *search space function* f and every input set Δ:

(2.6) $$\Delta \subseteq f(\Delta),$$

and the virtual space must be included in the search space, that is, for every *virtual space function* g, any search space function f compatible with g must satisfy:

(2.7) $$g(\Delta) \subseteq f(\Delta).$$

Finally, any virtual space function g must satisfy

(2.8) $$\mathtt{at}(\Delta) \subseteq g(\Delta)$$

where at is the function that, when applied to Δ, yields all the atomic subformulae of the formulae contained in it.

The functions used to delimit the search space and the virtual space are important parameters of an approximation system. We have seen that for the 0-depth layer, the specification of the virtual space function is irrelevant, since there is no use of virtual information, and the function sub is sufficient to delimit the search space. In general, for $k > 0$, the choice of specific functions to yield suitable values of these two parameters for each particular problem is the result of decisions that are conveniently made by the system designer, depending on the intended application. Such decisions affect the deductive power of each given k-depth approximation, and so the "speed" at which the approximation process converges to full Boolean logic. As we shall see, important special cases that are sufficient to define an approximation system for classical propositional logic in the sense of Section 1.1.2 are:

2.1. WEAK DEPTH-BOUNDED APPROXIMATIONS

- the search space function $f = \text{sub}$ and the virtual space function $g = \text{at}, \, ;$

- $f = g = \text{sub}$; in this case, the search space and the virtual space are identical.

Although both choices are suitable to define an approximation system that converges to classical propositional logic, the second choice provides much faster approximations with a negligible overhead. However, other choices may deliver even faster, yet tractable, approximations and the optimal choice depends on the application problem under consideration.

To summarize, let $\mathcal{H}$ be the set of all unary operations h on finite sets of $\mathcal{L}$-formulae such that, for all Δ:

1. $\Gamma \subseteq \Delta \implies h(\Gamma) \subseteq h(\Delta)$;

2. $\text{sub}(h(\Delta)) = h(\Delta)$.

The operations in $\mathcal{H}$ are partially ordered by the relation $\trianglelefteq$ such that $h_1 \trianglelefteq h_2$ if and only if, for every finite Δ, $h_1(\Delta) \subseteq h_2(\Delta)$. We say that h is *polynomially bounded* when for every Δ, $|h(\Delta)| \leq p(|\Delta|)$ for a fixed polynomial p, where $|\Gamma|$ denotes the *size* of Γ, that is the *total number of symbols* occurring in the formulae in Γ.

In order to define hierarchies of paracomplete logical systems that converge to classical propositional logic, suitable virtual space functions are those in the set:

(2.9) $$\mathcal{V} = \{g \in \mathcal{H} \mid \text{at} \trianglelefteq g\}.$$

The choice of a suitable search space function depends on the choice of the virtual space function, because the search space must include the virtual space. Moreover, the search space must include the input set. So, for each choice of a virtual space function $g \in \mathcal{V}$, suitable choices of the search space function must belong to the set:

(2.10) $$\mathcal{S}^g = \{f \in \mathcal{H} \mid g \trianglelefteq f, \text{sub} \trianglelefteq f\}.$$

Typical examples of operations that are both in $\mathcal{V}$ and in $\mathcal{S}^g$ are:

- the operation sub;

- the operations sub_k defined as follows (for a language $\mathcal{L}$ with logical operators $\star_1, \ldots, \star_n$):

    - $\text{sub}_0(\Delta) = \text{sub}(\Delta)$
    - for $k > 0$, $\text{sub}_k(\Delta) = \text{sub}_{k-1}(\Delta) \cup \bigcup_{1 \leq i \leq n} \{\star_i(A_1, \ldots, A_{a_i}) \mid A_1, \ldots A_{a_i} \in \text{sub}_{k-1}(\Delta)\}$,

    where a_i is the arity of $\star_i$;

- the operation at^* such that $\text{at}^*(\Delta)$ is the set of all formulae of $\mathcal{L}$ that can be built up from $\text{at}(\Delta)$;

- the constant operation F such that $\text{F}(\Delta) = F(\mathcal{L})$.

The operations sub, and sub_k are polynomially bounded, while F and at^* are not. Typical operations g that are in $\mathcal{V}$, but not in $\mathcal{S}^g$ are

- the operation at;

- the operation at^k such that $\text{at}^k(\Delta)$ is the set of all formulae of $\mathcal{L}$ of logical complexity at most k that can be built up from $\text{at}(\Delta)$,

which are both polynomially bounded. Observe also that, according to our definitions, the smallest possible search space is yielded by the function sub and the greatest virtual space compatible with it is given by the same function sub. Depending on the application context, one may want to bound the virtual space to a set that does not coincide with the set of subformulae of the input formulae, or even to one that properly includes it, because the resulting approximations can be deductively more powerful.

Given an arbitrary virtual space function g, a typical function in $\mathcal{S}^g$ is:

$$(2.11) \qquad f^g(\Delta) = \text{sub}(\Delta) \cup g(\Delta),$$

where Δ is the input set, which is polynomially bounded whenever g is.[3] This is the least search space function in $\mathcal{S}^g$ (with respect to the ordering $\trianglelefteq$). We shall call f^g the *canonical search space function* for the virtual space function g, and $f^g(\Delta)$ the *canonical search space* for g, where Δ is the input set.

---

[3]Observe that, in general, it is not always the case that $\text{sub}(\Delta) \subseteq g(\Delta)$ or viceversa.

2.1. WEAK DEPTH-BOUNDED APPROXIMATIONS

Setting the search space to $F(\Delta)$ or $\text{at}^*(\Delta)$ — namely to the set of all formulae of the language or to the set of all formulae that can be built up from the atomic formulae occurring in Δ — does not prevent us from defining an approximation system whenever we can prove that, for *each k-depth approximation* we can restrict our attention to a polynomially bounded subset, such as the canonical search space, with no loss of deductive power.[4] On the other hand such functions would not be suitable to delimit the virtual space. For example, as we shall see, setting the virtual space to $F(\Delta)$ would not yield approximations at all, since all classical inferences and tautologies could be proven at depth 1.

2.1.3 Weak k-depth approximations

The kind of depth-bounded approximations considered below satisfy the conditions mentioned at the end of the previous section: we can restrict ourselves to a search space that is polynomially bounded (the canonical search space), with no loss of deductive power for each depth k and each given virtual space function g, provided that the latter is polynomially bounded. Indeed, setting the search space to any superset of $f^g(\Delta)$ will not increase the deductive power of the approximations. For this reason we can omit the specification of the search space when talking about such depth-bounded approximations that we shall call *weak*. The situation is different in the case of the strong depth-bounded approximations that will be introduced in Section 2.2.

Suppose that, for a given input set, the the virtual space is given by a $\mathcal{L}$-domain Λ and the search space by a $\mathcal{L}$-domain Δ that includes Λ. The *k-depth forcing* relation for the pair (Λ, Δ) is defined as follows:

Definition 2.1.3. *Let Λ be a $\mathcal{L}$-domain, v a shallow information state over any $\mathcal{L}$-domain*[5]*, $\Delta \supseteq \Lambda$ and $x \in \{0,1\}$.*

- $v \Vdash_0^\Lambda \langle A, x \rangle \iff v(A) = x;$

- $v \Vdash_{k+1}^\Lambda \langle A, x \rangle$ *if, for some $B \in \Lambda$, $w \vDash_k^\Lambda \langle A, x \rangle$ for all $w \sqsupseteq v$ such that $w(B) \neq \bot$.*

---

[4]This is the typical content of a *subformula theorem* when $f = \text{sub}$.
[5]See Definitions 1.4.19 and 1.4.20.

Remark 2.1.4. Note that $\Vdash_0^\Lambda$ should not be confused with $\Vdash_{\text{SCP}}$ and is more powerful. For example, consider a valuation v such that $v(A \vee B) = 1$, $v(A) = 0$, $v(B \to C) = 1$ and $v(B) = v(C) = \bot$. The module α whose top formula is $A \vee B$ is unstable and $\alpha \Vdash_{\text{SCP}} \langle B, 1 \rangle$, therefore $v \Vdash_{\text{SCP}} \langle B, 1 \rangle$, in accordance with our definitions (see p. 31 above). On the other hand, the module β whose top formula is $B \to C$ is stable. So, $\beta \nVdash_{\text{SCP}} \langle C, 1 \rangle$ and $v \nVdash_{\text{SCP}} \langle C, 1 \rangle$. However $v \Vdash_0^\Lambda \langle C, 1 \rangle$, because shallow information states are closed under SCP. So, $\Vdash_{\text{SCP}}$ corresponds to a single and local application of SCP, while $\Vdash_0^\Lambda$ corresponds to any chain of such applications (see footnote 1).

Definition 2.1.5. Given a virtual space function v, a weak approximation of type g is a relation between finite sets of $\mathcal{L}$-formulae and formulae such that for all $\Gamma, A, f \in \mathcal{S}^g$:

- $\Gamma \models_k^g A$ if and only if $v \Vdash_k^\Lambda \langle A, 1 \rangle$, for all shallow information states v over Δ such that v verifies all the formulae in Γ, where

$$\Lambda = g(\Gamma \cup \{A\}) \text{ and } \Delta = f(\Gamma \cup \{A\});$$

- $\Gamma \models_k^g$ if and only if there is no shallow information state v over Δ that verifies all the formulae in Γ, such that $v \Vdash_k^\Lambda \langle A, 1 \rangle$, where

$$\Lambda = g(\Gamma) \text{ and } \Delta = f(\Gamma).$$

It immediately follows from the above definitions that:

Proposition 2.1.6. For all $g, g' \in \mathcal{V}$ and all $k, m \in \mathbb{N}$

$$\text{if } g \trianglelefteq g', \text{ and } k \leq k', \text{ then } \models_k^g \subseteq \models_{k'}^{g'}.$$

Observe that every shallow information state over Δ such that $v(p) \neq \bot$ for every atomic p occurring in Δ is a Boolean valuation of Δ, hence:

Proposition 2.1.7. Let $\models_C$ be the consequence relation of classical propositional logic. Then:

$$\bigcup_{k=0}^{\infty} \models_k^g \; = \; \models_C$$

for every $g \in \mathcal{V}$.

2.1. WEAK DEPTH-BOUNDED APPROXIMATIONS

As anticipated, not all virtual space functions $\models_k^g$ are adequate to generate a hierarchy of tractable depth-bounded approximations to Boolean Logic. For example, if the virtual space function is F, i.e., the virtual space is the set of all formulae of $\mathcal{L}$, then:

Proposition 2.1.8. $\models_1^F$ *is the consequence relation of classical propositional logic.*

To see this, consider that $\emptyset \models_1^F A$ for every classical tautology A. It is not difficult to check, using the sound inference rules of Table 1.5, that for any axiom of any of the standard Hilbert-style axiomatic systems for classical propositional logic, its negation is 0-depth inconsistent (see Remark 2.1.2), and so there is no 0-depth information state that falsifies A at depth 0. Moreover, $\models_1^F$ is closed under Modus Ponens (because $\models_0$ is) and, therefore, every classical tautology is a tautology also for $\models_1^F$.

2.1.4 C-intelim tableaux

Given an S-formula φ, its *unsigned part* φ^u is the unsigned formula that results by removing the sign.[6] Given a set X of S-formulae, $X^u = \{\varphi^u \mid \varphi \in X\}$. The deducibility relation $\vdash_k^g$ (for S-formulae) that corresponds to the weak k-depth consequence relation $\models_k^g$ can be defined in a straightforward way, starting from $\vdash_0$:

Definition 2.1.9. *For all $g \in \mathcal{V}$, and all X, ψ:*

1. (a) $X \vdash_0^g \psi$ *if and only if* $X \vdash_0 \psi$;
   (b) $X \vdash_0^g$ *if and only if* $X \vdash_0$.
2. (a) $X \vdash_{k+1}^g \psi$ *if and only if there is* $A \in g(X^u \cup \{\psi^u\})$, *such that* $X \cup \{T\,A\} \vdash_k^g \psi$ *and* $X \cup \{F\,A\} \vdash_k^g \psi$;
   (b) $X \vdash_{k+1}^g$ *if and only if there is* $A \in g(X^u \cup \{\psi^u\})$, *such that* $X \cup \{T\,A\} \vdash_k^g$ *and* $X \cup \{F\,A\} \vdash_k^g$.

The version for unsigned formulae is obtained by replacing X with $\text{Uns}(X)$ and φ with $\text{Uns}(\varphi)$. In what follows, definitions and propositions

---

[6]This is not to be confused with the *unsigned version* of φ, denoted by $\text{Uns}(\varphi)$ defined above on p. 41. The unsigned part of $F\,A$ is A, while the unsigned version of $F\,A$ is $\neg A$.

will mostly be in terms of S-formulae and we leave it to the reader to work out the corresponding versions for unsigned formulae.

The above definition is equivalent to augmenting the intelim rules with the following depth-increasing *deduction rule*:

$$\frac{X \cup \{TA\} \vdash^g_k \varphi \quad X \cup \{FA\} \vdash^g_k \varphi}{X \vdash^g_{k+1} \varphi.}$$

By analogy with $\models^g_k$ we say that the depth-bounded deducibility relation $\vdash^g_k$ is of type g.

Given Definitions 2.1.5, 2.1.9 and Proposition 1.4.3, it is far from surprising that:

Proposition 2.1.10. *For every finite set X of signed formulae and every signed formula φ,*

- $X \vdash^g_k \varphi$ *if and only if* $X \models^g_k \varphi$,

- $X \vdash^g_k$ *if and only if* $X \models^g_k$.

Again

As expected, $\vdash^g_k$ is monotonic on g and on k:

$$\text{if } g \trianglelefteq g', \text{ and } k \leq k', \text{ then } \vdash^g_k \subseteq \vdash^{g'}_{k'}.$$

Deductions of $X \vdash^g_k \varphi$ ($\text{Uns}(X) \vdash^g_k \text{Uns}(\varphi)$) can be presented in the format of a tree whose branches are intelim sequences and virtual information is introduced by a *branching rule* that splits an intelim sequence as follows (depending on whether it is used for signed or unsigned formulae):

(RB)

$$\boxed{\text{For all } A \in g(X^u \cup \{\varphi^u\})} \qquad \boxed{\text{For all } A \in g(\text{Uns}(X) \cup \{\text{Uns}(\varphi)\})}$$

$$\vdots \qquad\qquad\qquad\qquad \vdots$$
$$\diagup \;\diagdown \qquad\qquad\qquad \diagup \;\diagdown$$
$$TA \quad FA \qquad\qquad\quad A \quad \neg A$$

We call the formulae TA and FA (A and $\neg A$) *virtual assumptions*, while the assumptions in X ($\text{Uns}(X)$) are called *actual assumptions*. We also say that RB has been *applied to* A and its application is *g-analytic* (in the syntactic sense of "analytic").

Proofs and refutations in this system take the form of downward-growing trees of S-formulae (or of unsigned formulae),[7] similar to Smullyan's Analytic Tableaux (Smullyan, 1968) except that they can serve as proofs as well as refutations, because of the rôle played by the introduction rules. For this reason we shall call them *C-intelim tableaux* (of type g when the applications of RB are all g-analytic). On the other hand, they also bear a resemblance with natural deduction because of the intelim rules and of their "harmony" (Lemma 1.4.9).

When proofs are presented in this format, $X \vdash^g_k \varphi$ if and only if there is a tree $\mathcal{T}$ such that (i) each branch of $\mathcal{T}$ contains at most k virtual assumptions introduced via the branching rule RB (so $\mathcal{T}$ contains at most 2^k branches), (ii) each branch is an intelim sequence for $X \cup Y$, where Y is the set of all the virtual assumptions on that branch; (iii) a branch is *closed* if it is a closed intelim sequence, i.e. it contains both φ and its conjugate $\overline{\varphi}$ for some signed formula φ and (iii) φ occurs in all *open* branches (namely, those that are not closed). Note that (iii) implies that a C-intelim tableau such that all its branches are closed represents a "proof" of *any* S-formula φ from its actual assumptions. We call such "proofs" *improper* (see Definition 1.4.7).

When a branch is closed we mark it with $\times$ as in Smullyan's tableaux. In this context we shall sometimes say that "the closure rule" has been applied to the two conjugate S-formulae or unsigned formulae, albeit strictly speaking no inference rule is applied: $\times$ is not part of the language and is only an "inconsistency detector". For unbounded k and $g = \text{sub}$, C-intelim tableaux are a proof system for full classical propositional logic that enjoys the *generalized subformula property* (see Proposition 2.1.19 below): every S-formula ψ occurring in a C-intelim proof of φ from X (in a refutation of X) is such that its unsigned part ψ^u is a subformula of some formula in $X^u \cup \varphi^u$ (of X^u). As anticipated, it may make sense to consider larger virtual spaces than those yielded by sub, in that they may deliver more powerful subsystems. On the other hand, by removing the restriction on A, the rule RB allows also for representing proofs with unrestricted "lemmas": any formula A such that FA can be refuted on the basis of the premises,

---

[7]This is a sloppy way of saying that they are trees whose nodes, except possibly the root, are labelled with S-formulae (with unsigned formulae).

can be used as a lemma. An extreme case (which is useless from the point of view of designing an approximation system) is when the virtual space is the set of all $\mathcal{L}$-formulae. Proposition 2.1.8 implies that, when $f = g = \mathrm{F}$, C-intelim tableaux of depth 1 can efficiently simulate unrestricted Frege systems,[8] which are among the most powerful in the p-simulation hierarchy (Urquhart, 1995).[9] Less uncompromising, polynomially bounded, search space functions $f \triangleright \mathrm{sub}$ may turn out to be useful in some applications.

Note that, unlike intelim sequences, for depth greater than 0, C-intelim tableaux may based on the *empty set of assumptions*, that is the root of the tree is not labelled by any signed or unsigned formula and tree starts with an application of the branching rule RB.

Examples of C-intelim tableau proofs are shown in Figures 2.1 and 2.2. In their construction we have used *the unsigned intelim rules* to illustrate their use. Moreover, we have used schematic letters $A, B, \ldots$, to stress the fact that proofs are still correct for any substitution of the schematic letters, provided that the restriction on the virtual space is satisfied. The three ones in Figure 2.1 are based on non-empty set of assumptions. The first and the third one have depth 1, while the one in the middle has depth 2. Closed branches are marked with the symbol $\times$.

It is customary to list all the actual assumptions at the beginning, starting from the root. Proofs from the empty set of assumptions are represented by trees with an unlabelled root as illustrated in Figure 2.2. The first example shows how the introduction rules can be used to simulate the truth-table method. The second example shows how to represent a typical pattern of proof ex-absurdo. C-intelim tableaux can be naturally used as a *refutation system*, like resolution or semantic tableaux, as well as a system of direct proof. In fact, if we disallow the introduction rules, we obtain the system KE (D'Agostino & Mondadori, 1994), which is a variant of Smullyan's se-

---

[8]This is the standard terminology for Hilbert style axiomatic systems in the literature on the computational complexity of proof systems.

[9]On the connection between the rule RB and the cut rule of Gentzen's sequent calculus, as well as on the advantages of similar *cut-based* formalizations of classical logic, see D'Agostino (1990, 1992); D'Agostino & Mondadori (1994); D'Agostino (1999). We call a formalization of classical logic *cut-based* if it contains some form of the classical cut rule that is *not redundant*, but can be "tamed", that is, can be restricted to applications where the cut formula is a subformula of the input formulae. The fact that cut is not redundant is essential in that this implies that this rule is applied only when needed.

2.1. WEAK DEPTH-BOUNDED APPROXIMATIONS

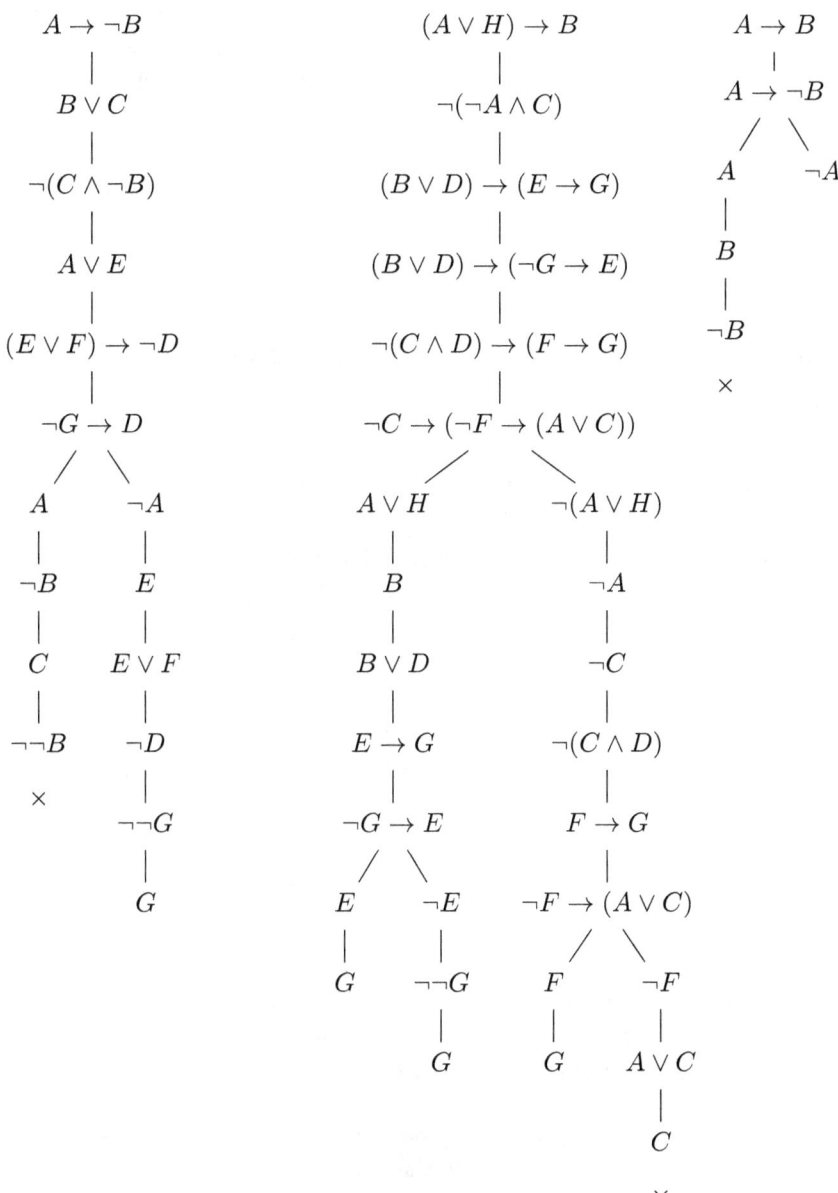

Figure 2.1: C-intelim tableaux. Each branch is an intelim sequence.

80   CHAPTER 2. DEPTH-BOUNDED DEDUCTION

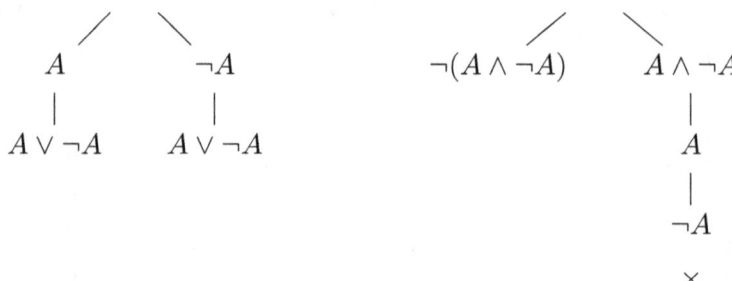

Figure 2.2: C-intelim tableau proofs from the empty set of assumptions.

mantic tableaux, but essentially more efficient.[10] On the other hand if we disallow the elimination rules we obtain the system KI (Mondadori, 1988c, 1995; D'Agostino, 1999), which can be regarded as a proof-theoretical version of the truth-table method (but essentially more efficient).[11] Using both introduction and elimination rules allows for shorter[12] deductions that correspond more closely to actual human reasoning and require fewer applications of the discharge rule RB.

2.1.5 Normal and quasi-normal tableaux

A *path* in a C-intelim tableau $\mathcal{T}$ is a finite sequence of nodes such that the first node is the root of $\mathcal{T}$ and each subsequent node occurs immediately below the previous one (so, a branch is a maximal path). A path is *closed* if it contains occurrences of both $T B$ and $F B$ for some formula B.

Definition 2.1.11. *A C-intelim tableau is* non-redundant *if (i) no branch of $\mathcal{T}$ contains more than one occurrence of the same formula, and (ii) no branch of $\mathcal{T}$ properly contains a closed path.*

---

[10] As shown in (D'Agostino & Mondadori, 1994), KE can p-simulate analytic tableaux but analytic tableaux cannot p-simulate KE. In fact, analytic tableaux cannot even p-simulate the truth-tables (D'Agostino (1992)).

[11] The truth-table method cannot p-simulate KI (Mondadori, 1995).

[12] But not *essentially* shorter, for both KI and KE can p-simulate C-intelim (D'Agostino, 1999).

2.1. WEAK DEPTH-BOUNDED APPROXIMATIONS 81

Since every branch of a non-redundant C-intelim tableaux is a non-redundant intelim sequence, by Lemma 1.4.11 non-redundant tableaux contain no detours (Definition 1.4.8).

Definition 2.1.12. *An occurrence of a formula in a C-intelim tableau is* idle *if (i) it is not the terminal node of the branch and (ii) it is not used as premise of an application of an intelim rule or of the "closure rule" in the subtree descending from it.*

Definition 2.1.13. *A C-intelim tableau* $\mathcal{T}$ *is* strongly non-redundant *if it is non-redundant and contains no idle occurrences of formulae.*

Proposition 2.1.14. *Every C-intelim tableau can be turned into a strongly non-redundant one.*

The procedure is quite simple:

1. if a branch properly contains a closed path, remove the subtree descending from the last node of the closed path;

2. remove all idle S-formulae and all repetitions of the same S-formula from each branch; if the idle or repeated occurrences of an S-formula is a virtual assumption introduced by an application of RB, remove also the whole subtree descending from the sibling node; do this until there are no more idle or repeated S-formulae.

The elimination of idle or repeated occurrences of a formula in a branch may turn some previously used occurrences of formulae into idle ones; but at each reduction step the size of the tree decreases, and so the procedure terminates in a number of steps that is linear in the size of the initial C-intelim tree.

Example 2.1.15. *An example is illustrated in Figure 2.3 for C-intelim trees based on the* unsigned rules *(the actual assumptions are marked with "ⓐ"). In the leftmost tree the occurrence of $\neg D$ is not idle, because it is used as premise of modus ponens to obtain $\neg E$ at the end of the left branch. On the other hand, the virtual assumption on the right branch is idle, so we have to remove it together with the whole subtree descending from the sibling node.*

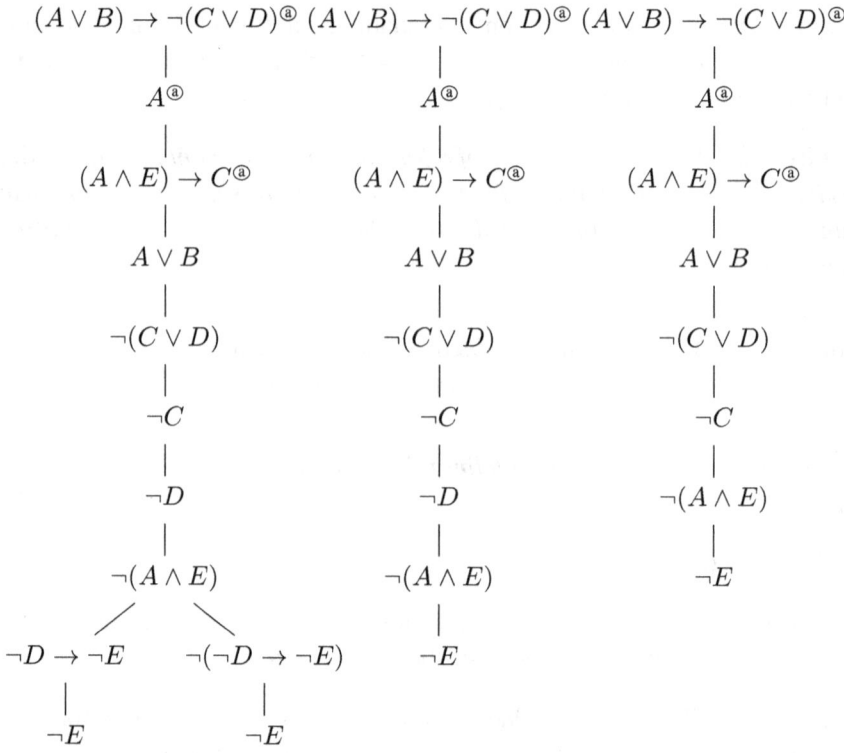

Figure 2.3: Turning a C-intelim tableau into a strongly non-redundant one.

Thus, we obtain the intelim sequence in the middle. Now, the occurrence of $\neg D$ has turned idle, so we must just remove it to obtain the rightmost non-redundant C-intelim tableau.

Definition 2.1.16. *A C-intelim proof or refutation $\mathcal{T}$ is g-normal if (i) $\mathcal{T}$ is strongly non-redundant and (ii) all applications of RB in $\mathcal{T}$ are g-analytic for some $g \in \mathcal{V}$ such that $|g(\Delta)|$ is bounded above by a polynomial in $|\Delta|$.[13] It is* atomically normal *if it is normal and all applications or RB are atomic.*

When every application of RB is *analytic* (i.e., g-analytic with $g = \text{sub}$) we just say that $\mathcal{T}$ is *normal*. If $\text{sub} \triangleleft g$ (i.e., $\text{sub} \trianglelefteq g$ and $g \neq \text{sub}$), we

---

[13] Recall that $|\Gamma|$ is the total number of occurrences of symbols in the formulae in Γ.

2.1. WEAK DEPTH-BOUNDED APPROXIMATIONS

say that $\mathcal{T}$ is *quasi-normal*. We stress that quasi-normality is an important generalization of normality in the context of depth-bounded approximations, because in some cases the minimum depth of a quasi-normal proof is smaller than the minimum depth of a normal proof. In general, whenever $g_1 \triangleleft g_2$ the minimum depth of a g_1-normal proof may be greater than that of a g_2-normal proof.

A proof of the next lemma is sketched in Section A.5).

Lemma 2.1.17. *Given any $g \in \mathcal{V}$, every k-depth intelim proof $\mathcal{T}$ of φ from X (k-depth intelim refutation of X) can be transformed into a $k + j$-depth intelim proof $\mathcal{T}'$ of φ from X (intelim refutation $\mathcal{T}'$ of X), for some $j \geq 0$, such that every application of RB in $\mathcal{T}$ is g-analytic or atomic.*

Note, the the "or" in the above lemma is inclusive. The result of the transformation may or may not contain applications that are not g-analytic. But applications that are not g-analytic (if any) must be atomic. In fact, the transformations used in the proof of the above proposition show that every C-intelim tree can be transformed into an equivalent one in which all the RB-formulae are *atomic*. So, in principle, we could reformulate the notion of intelim tree in such a way that RB is applied only to atomic formulae without loss of completeness. However, each application of these transformations increases the depth of the tree, so that the property of being a k-depth intelim tree (for a fixed k) is not necessarily preserved under uniform substitutions of the atomic formulae occurring in the tree without increasing the depth of the tree. On the other hand, if we require that the notion of C-intelim tree be restricted so as to permit only *analytic* applications of RB (with $f = \text{sub}$), the property of being an intelim tree is indeed invariant under uniform substitutions, with no increase in the depth of the tree.

By Lemma 2.1.17, every C-intelim proof $\mathcal{T}$ of φ from X (every C-intelim refutation of X) can be transformed into a C-intelim proof $\mathcal{T}'$ of φ from X (C-intelim refutation of X) such that all the applications of RB are g-analytic or atomic. Observe that, by Proposition 2.1.14 $\mathcal{T}'$ can be transformed into a strongly non-redundant C-intelim proof $\mathcal{T}''$ of φ from X (C-intelim refutation of X). It is not difficult to show that a strongly non-redundant C-intelim tableau cannot contain applications of RB to atomic S-formulae that are not signed subformulae of the input formulae. So all applications of RB are g-analytic. It follows that:

Proposition 2.1.18. *Given any $g \in \mathcal{V}$, every C-intelim proof $\mathcal{T}$ of φ from X (C-intelim refutation of X) can be transformed into a g-normal one.*

The following proposition states the generalization of the SFP (GSFP) for intelim proofs and refutations:

Proposition 2.1.19 (Generalized SFP). *For every $g \in \mathcal{V}$, if $\mathcal{T}$ is a g-normal proof of φ from X, or a g-normal refutation of X, then for every S-formula ψ occurring in $\mathcal{T}$, ψ^u belongs to the canonical search space, i.e.:*

$$\psi^u \in g(X^u \cup \{\varphi^u\}) \cup \texttt{sub}(X^u \cup \{\varphi^u\})$$

if $\mathcal{T}$ is a proof of φ from X, or

$$\varphi^u \in g(X^u) \cup \{\texttt{sub}(X^u)\}$$

if $\mathcal{T}$ is a refutation of X.

In other words, every S-formula that occurs in the proof is a signed subformula either of a premise or of the conclusion. A detailed proof can be adapted from that of Proposition 1.4.15 (see Section A.2).

The GSFP implies that in the search for g-normal C-intelim proofs or refutations we can restrict ourselves to the *canonical search space* for g (see p. 72). In the special case in which $g = \texttt{sub}$ we obtain the usual SFP. Observe that, since 0-depth C-intelim tableaux have no virtual assumptions, every g-normal 0-depth proof or refutation is normal and has the SFP. The above propositions can be adapted to trees of unsigned formulae in the obvious way and, when dealing with such trees, the SFP is replaced, as before, by the WSFP (see p. 53).

Let $\vdash_\infty^g$ be the unbounded deducibility relation defined as follows: $X \vdash_\infty^g \varphi$ ($X \vdash_\infty^g$) if there is a g-normal intelim proof of φ from X (a g-normal refutation of X) of any depth $k \in \mathbb{N}$. Proposition 2.1.18 guarantees that $\vdash_\infty^g$ is complete for classical propositional logic. This implies, among other things, that the application of the introduction rules can be *goal-oriented* in the sense clarified by the following:

Proposition 2.1.20. *Let $\mathcal{T}$ be a g-normal proof of φ from X (refutation of X) and let $\psi_1, \ldots, \psi_n$ ($n > 1$) be a maximal sequence of formulae occurring in a branch of $\mathcal{T}$ such that, for every $i = 1, \ldots, n$, ψ_i is the conclusion of an application of an introduction rule to previous S-formulae in the sequence. Then one of the following holds true:*

2.1. WEAK DEPTH-BOUNDED APPROXIMATIONS

1. ψ_n is the minor premise of an elimination;

2. $\psi_n = \varphi$;

3. ψ_n is one of the two premises of an application of the closure rule.

A proof is sketched in Section A.4.

It is easy to see, by inspection of the intelim rules, that there is no *need* to use a sequence of introduction rules in order to obtain the conjugate of an S-formula occurring in the same branch, because the branch can be closed by applying elimination rules. For example, if $T\,A \vee B$, $F\,A$ and $F\,B$ occur in the same branch, this can be closed by introducing $F\,A \vee B$ or by eliminating $T\,A \vee B$. eliminating $T\,A \vee B$.

Hence, the search for a proof or a refutation can be governed by a procedure that is informally described by the following four general rules for expanding an intelim tree:

1. stop expanding a branch whenever it is closed;

2. give priority to the elimination rules;

3. apply (suitable sequences of) introduction rules only to obtain either the conclusion of the proof, or a minor premise that is needed for an elimination;

4. apply the branching rule RB to an open branch only when instructions 2 and 3 fail.

The choice of the RB-formula in the last instruction depends on the operation g that defines the virtual space. When $g = \mathtt{sub}$, one can always, without loss of completeness, choose as RB-formula some subformula of the assumptions or of the conclusion that does not already occur in the branch. This is the procedure that has been followed in the construction of the trees in Fig. 2.1. If we apply these rules mechanically, the resulting intelim proof would contain no detours, but it may still contain idle S-formulae. Then, to obtain a g-normal proof or refutation, it is sufficient to remove them from the tree.

Following the decision procedure of Section A.3 (in the Appendix), once the search space is structured as a *subformula graph*, the order of application of the intelim rules in each branch is immaterial and only g-normal proofs or refutations are generated (modulo occurrences of idle S-formulae).

From Propositions 1.4.22 and 2.1.18, it follows that, for each $g \in \mathcal{V}$ and each fixed k, $\vdash^g_k$ admits of a feasible decision procedure (since g is assumed to be polynomially bounded).

Proposition 2.1.21. *For each $g \in \mathcal{V}$ and each $k \in \mathbb{N}$, whether or not $X \vdash^g_k \varphi$ ($X \vdash^g_k$), can be decided in polynomial time.*

An algorithm for $\vdash^g_k$ can obtained by adapting the Expand sub-routine in Section A.3 as follows. First, the initial set Γ must be taken to include all the formulae of maximal logical complexity in the canonical search space.. Second, we need to accommodate applications of the RB rule. When an intelim graph G is completed by running the 0-depth algorithm and there are yet undefined formulae, choose one of such formulae, say A, and split G into two sibling intelim graphs G_1 and G_2 in which A is labelled, respectively, with "1" and "0", while all the other formulae are labelled as in G. Next, complete G_1 and G_2 and iterate the procedure on each of them. The final result is binary tree of height k whose nodes are intelim graphs and the procedure terminates with output **true** if the 0-depth algorithm returns **true** for each of them. Considering al possible choices of PB-formulae, the algorithm generates a forest of such trees and returns **true** if Expand returns **true** for at least one of them.

When $g \trianglelefteq$ sub, the complexity of the decision problem is $O(n^{k+2})$, where n is the total number of occurrences of symbols in $X \cup \{\varphi\}$ (in X). A more accurate analysis shows that the procedure terminates in a number of steps less than or equal to

$$2^k \cdot \binom{n}{k} O(n^2),$$

so that the upper bound when $k = n$ is just exponential (and not n^n as the cruder analysis may suggest). In general, for a polynomially bounded g, the complexity is $O(p(n)^{k+2})$ where p is a polynomial depending on g.

2.2 Strong depth-bounded approximations

2.2.1 Depth-bounded information states

The characterization of stronger consequence relations that satisfy some form of Cut (i.e., "local transitivity", see p. 8) prompts us to introduce a more general notion of k-depth information state.

Definition 2.2.1. *For all $\mathcal{L}$-domains Δ and all Λ, such that $\Lambda \subseteq \Delta$, the set $\mathbf{S}\binom{\Delta}{\Lambda}_k$ is defined as follows:*

1. $\mathbf{S}\binom{\Delta}{\Lambda}_k$ *is the set of all shallow information states v over Δ such that v is closed under the following condition:*

$$v \Vdash_k^\Lambda \langle A, x \rangle \Longrightarrow v(A) = x.$$

Elements of $\mathbf{S}\binom{\Delta}{\Lambda}_k$ are called *k-depth information states of type* $\binom{\Delta}{\Lambda}$. Observe that, for every $\mathcal{L}$-domain Δ and every $\mathcal{L}$-domain $\Lambda \subseteq \Delta$,

$$\mathbf{S}\binom{\Delta}{\emptyset}_0 = \mathbf{S}\binom{\Delta}{\Lambda}_0 = \mathbf{S}\binom{\Delta}{\emptyset}_k$$

is the set of all shallow information states over Δ.

For strong depth-bounded approximations, as will be apparent in the next section, we may be unable to define a canonical search space and so we need to specify not only the virtual space, but also the search space.

Definition 2.2.2. *For all $g \in \mathcal{V}$ and all $f \in \mathcal{S}^g$ the strong k-depth approximation of type $\langle f, g \rangle$ is the relation $^+\models_k^{f,g}$ defined as follows:*

- $\Gamma\ ^+\models_k^{f,g} A$ *iff A is verified by all k-depth information states $v \in \mathbf{S}\binom{\Delta}{\Lambda}_k$ such that v verifies all the formulae in Γ, where:*

$$\Delta = f(\Gamma \cup \{A\}) \text{ and } \Lambda = g(\Gamma \cup \{A\}).$$

- $\Gamma\ ^+\models_k^{f,g}$ *if there is no k-depth information state $v \in \mathbf{S}\binom{\Delta}{\Lambda}_k$ such that v verifies all the formulae in Γ, where:*

$$\Delta = f(\Gamma) \text{ and } \Lambda = g(\Gamma).$$

Proposition 2.2.3. *For all $g, g' \in \mathcal{V}$, all $f, f' \in \mathcal{S}^g$ and all $k, k' \in \mathbb{N}$,*

if $g \trianglelefteq g'$, $f \trianglelefteq f'$ and $k \leq k'$, then ${}^+\!\models_k^{f,g} \subseteq {}^+\!\models_k^{f',g'} {}^+\!\models_{k'}^{f,g}$.

We shall use the notation ${}^+\!\models_\infty^{f,g}$ to refer to the limit of the sequence of all k-depth approximations of type $\langle f, g \rangle$, that is, ${}^+\!\models_\infty^{f,g} =_{def} \bigcup_{k=0}^\infty {}^+\!\models_k^{f,g}$. Given Proposition 2.1.7 above, *a fortiori* it holds true that:

Proposition 2.2.4. *Let $\models_C$ be the consequence relation of classical propositional logic. Then:*

$$\bigcup_{k=0}^\infty {}^+\!\models_k^{f,g} = \models_C .$$

Not all strong k-depth approximations are Tarskian consequence relations, for they may or may not satisfy local transitivity, depending on their type.[14] Similarly, not all of them satisfy local substitution invariance, for example when the virtual space is delimited by the function at.

Although they may not satisfy unrestricted Cut, strong k-depth consequence relations always satisfy the following restricted version of it:

(Bounded Cut) For all $B \in f(\Gamma \cup \{A\})$,

$$\Gamma \,{}^+\!\models_k^{f,g} B \text{ and } \Gamma \cup \{B\} \,{}^+\!\models_k^{f,g} A \implies \Gamma \,{}^+\!\models_k^{f,g} A.$$

An important special case of Bounded Cut is the one in which $f = g = \mathsf{sub}$, and so

$$f(\Gamma \cup \{A\}) = g(\Gamma \cup \{A\}) = \mathsf{sub}(\Gamma \cup \{A\}),$$

in which case we speak of *Analytic Cut*. Whatever their type, strong depth-bounded approximations satisfy the requirements of Global Transitivity and Global Substitution Invariance (p. 8).

2.2.2 C-Intelim hyper-sequences

The proof-theory of strong depth-bounded approximations is best presented in terms of C-intelim *hyper-sequences*. We extend the logical language to include a symbol $\curlywedge$ that stands for a formula that is not verified by any

---
[14] See the requirements for an approximation system on p. 8.

2.2. STRONG DEPTH-BOUNDED APPROXIMATIONS

shallow information state. This move, strictly speaking, is not necessary, but avoids cumbersome definitions. We make the simplifying assumption that λ never occurs in the premises. Let us add to the C-intelim rules the following λ-rules:

$$\frac{\begin{array}{c}TA\\FA\end{array}}{T\lambda}\text{ RNC} \qquad \frac{T\lambda}{\varphi}\text{ XFQ.}$$

When dealing with unsigned formulae the rules take the following form:

$$\frac{\begin{array}{c}A\\\neg A\end{array}}{\lambda}\text{ RNC} \qquad \frac{\lambda}{A}\text{ XFQ.}$$

The rule RNC expresses (the informational version of) the Principle of Non-Contradiction discussed in Chapter 1. So, in the sequel, a closed C-intelim sequence will be represented by a sequence ending with $T\lambda$ (or with λ when using the unsigned version of the rules) and, for ease of expression, we shall call it a "proof" of $T\lambda$ (of λ). The rule XFQ expresses the *ex-contradictione quodlibet* principle that holds for classical logic as well as for its depth-bounded approximations (see p. 34).

Accordingly, we allow λ also to be a value of the metalinguistic variables for unsigned formulae and extend the notion of *intelim sequence* to allow for the application of the falsum rules.

Definition 2.2.5. *The notion of k-depth C-intelim hyper-sequence for X of type g is defined as follows:*

- *Every intelim sequence for X is a 0-depth C-intelim hyper-sequence for X of type g for all $g \in \mathcal{V}$;*

- *For $k > 0$, the sequence of S-formulae $\pi = \varphi_1, \ldots, \varphi_n$ is a k-depth C-intelim hyper-sequence for X of type g if for every $i = 1, \ldots, n$, either (i) φ_i is in X, or (ii) for some $B \in g(X^u \cup \{\varphi_i^u\})$, there is a $k-1$-depth C-intelim hyper-sequence π_1 of type g for*

$$\{\varphi_1, \ldots, \varphi_{i-1}, TB\}$$

and a $k-1$-depth C-intelim hyper-sequence π_2 of type g for

$$\{\varphi_1, \ldots, \varphi_{i-1}, F\,B\}$$

such that both π_1 and π_2 end with φ_i.

Definition 2.2.6. *A k-depth C-intelim* hyper-proof *of ψ from X of type g is a k-depth C-intelim hyper-sequence for X of type g such that its last formula is ψ. A k-depth C-intelim* hyper-refutation *of X of type g is a k-depth C-intelim hyper-proof of $T \curlywedge$ from X of type g.*

We write $X \mathrel{{}^+\vdash^g_k} \varphi$ to say that there is a k-depth hyper-proof of type g of φ from X. Observe that any k-depth intelim hyper-sequence (hyper-proof, hyper-refutation) of type g is also a k'-depth intelim hyper-sequence (hyper-proof, hyper-refutation) of type g' for all $k' \geq k$ and $g' \unrhd g$. So, it follows from our definitions that

(2.12) $\quad X \mathrel{{}^+\vdash^g_k} \varphi \implies X \mathrel{{}^+\vdash^{g'}_{k'}} \varphi$ for all $g' \unrhd g$ and all $k' \geq k$.

We call *C-intelim$^+$* the system based on the C-intelim rules in which proofs are (fully-blown) hyper-sequences, rather than trees. A proof in this system provides a representation not only of the main hyper-proof, but also of all the auxiliary sub-proofs needed to fully justify each step. To represent the introduction of virtual information in C-intelim$^+$, two parallel boxes may be opened that contain, respectively, the auxiliary $(k-1)$-depth hyper-sequences based on $X \cup \{T\,A\}$ and $X \cup \{F\,A\}$, for some $B \in g(X^u \cup \{\varphi^u\})$, where g is the virtual space function, as illustrated in Figure 2.4. The usual scoping rules for boxes are employed here: each S-formula occurring in a box can be used, as premise of a rule application, in every inner box and cannot be used in outer boxes. The S-formulae that can be used as premises in a box are said to be *active* for that box. The whole deduction should be regarded as being contained in a root box that contains all the others, although we shall not usually draw the borders of this most external box.

To illustrate the difference between C-intelim tableaux and C-intelim$^+$ proofs, consider the deduction of the conclusion $T\,t \vee u$ from the premises

$$T\,p \vee q, T\,p \to r \vee s, T\,q \to r \vee s, T\,r \to t, T\,s \to u.$$

2.2. STRONG DEPTH-BOUNDED APPROXIMATIONS

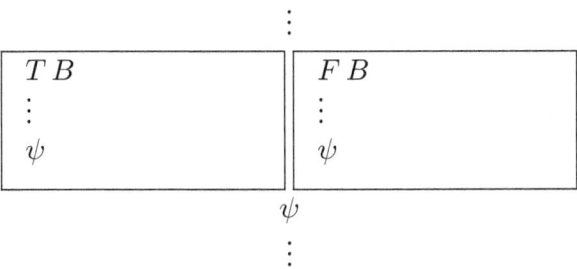

Figure 2.4: The format of RB in hyper-sequences.

The proof using the tree format represented in Figure 2.5 has depth 2 and the two subtrees generated by the nodes labelled with $T\, r \vee s$ are identical, while the proof using the C-intelim$^+$ format represented in Figure 2.6 has depth 1. Finally, in Figure 2.7 we illustrate the use of the $\curlywedge$-rules RNC and XFQ. We stress again that these rules are not essential and could be removed at the price of a more cumbersome format of the RB rule.

Each box in a C-intelim$^+$ proof is associated with the hyper-sequence $\varphi_1, \ldots, \varphi_n$ consisting of all the S-formulae that are active in that box, in the order in which they occur in the proof. This hyper-sequence is a hyper-proof of its last S-formula from $X \cup Y$, where Y is the set of all virtual assumptions that are active in the box. For example, in Figure 2.6, the hyper-sequence associated with the outermost box is the sequence of S-formulae occurring in lines 1–5, 9, 14 and is a hyper-proof of $T\, t \vee u$ from the premises in 1–5. The hyper-sequence associated with the lefthand box opened on line 6, consists of the S-formulae active in that box — namely those occurring in lines 1–5, 6,7 — and is a hyper-proof of $T\, r \vee s$ from the premises plus the virtual assumption $T\, p$. Given a C-intelim$^+$ proof Π, a *sub-proof* of Π consists in the C-intelim hyper-sequence associated with one of its boxes, including the outermost one, so that every proof Π is a subproof of itself. (We use the capital greek letter Π possibly with subscripts for C-intelim$^+$ proofs.[15])

By analogy with Definition 1.4.10, we say that *C-intelim hyper-sequence*

---

[15]Recall that in C-intelim hyper-sequences we allow for the application of the falsum as well as the intelim rules.

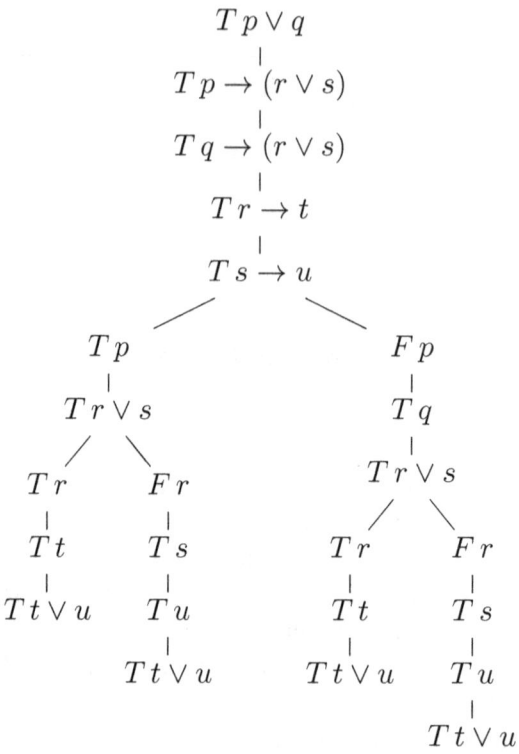

Figure 2.5: A C-intelim proof of depth 2 in tree format.

π is *non-redundant* when it satisfies the following conditions: (i) no subproof of π contains repeated S-formulae and (iii) no subproof of π properly contains a closed subsequence. It is *strongly non-redundant* if it is also true that (iii) no subproof of π contains idle occurrences of formulae and (iii) in no subproof of π an application of XFQ is followed by an application of an intelim rule. The rationale for condition (iii) is that, if $T \curlywedge$ occurs in a subproof of π, then the conclusion of any application of an intelim rule to the conclusion of XFQ can be obtained directly from $T \curlywedge$ by means of XFQ. A *C-intelim$^+$ proof* is *non-redundant* (strongly non-redundant) if all its subproofs are non-redundant (strongly non-redundant). Non-redundant

2.2. STRONG DEPTH-BOUNDED APPROXIMATIONS

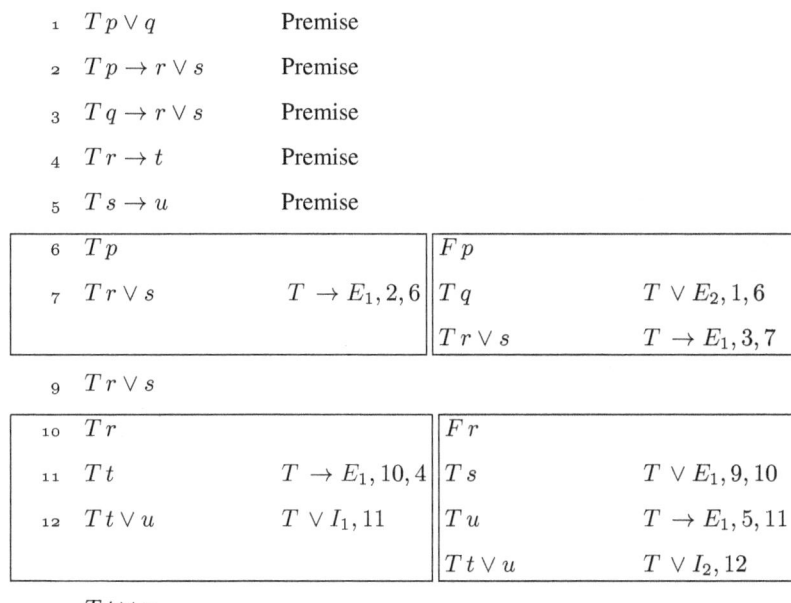

Figure 2.6: A C-intelim$^+$ proof of depth 1 of the same inference as in Fig. 2.5.

subproofs contain no detours (in the standard sense of Definition 1.4.8).

By definition each k-depth relation characterized by $^+\vdash_k^g$ satisfies unrestricted local transitivity, as well as reflexivity and monotonicity; moreover, if $g = $ sub satisfies also substitution invariance, and therefore is a Tarskian logic, i.e. a first class citizen in the domain of consequence relations. However, as anticipated in Section 2.1.3, in the case of strong depth-bounded approximations we must take care of the search space as well as of the virtual space. Indeed, for weak depth-bounded approximations, what we called the *canonical search space* is sufficient for completeness. To restrict proof-search to this space we only need to ensure that proofs or refutations are g-*normal*, that is strongly non-redundant and such that all the applications of RB in them are g-analytic.

In C-intelim$^+$ there is no guarantee that such proofs enjoy the GSFP (see Proposition 2.1.19), that is, in some cases a g-normal C-intelim$^+$ proof

94   CHAPTER 2. DEPTH-BOUNDED DEDUCTION

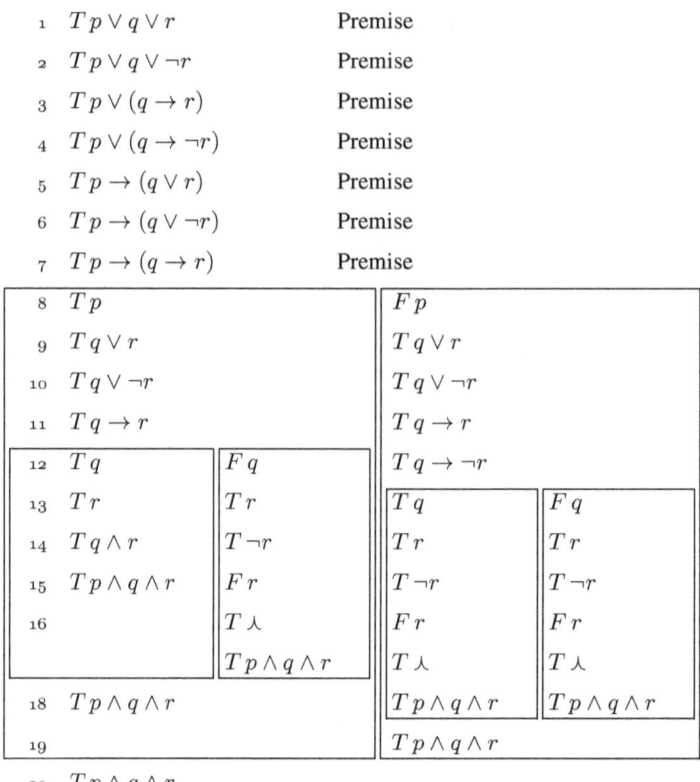

Figure 2.7: A C-intelim$^+$ proof of depth 2, making use of the λ-rules.

may contain occurrences of S-formulae that do not belong to the canonical search space. Therefore, bounding the virtual space is not sufficient for the GSFP. This is the main reason that motivated us to introduce a specification of the search space in the definition of strong depth-bounded approximations. Figure 2.8 shows a critical example of a 1-depth C-intelim$^+$ proof of type sub, so that the canonical search space is simply sub(Δ), but does not have the SFP, although it is strongly non-redundant and all applications of RB in it are syntactically analytic. In order to transform it into one with the SFP we need to increase its depth (in this case to depth 2). This counterexample raises the problem of the tractability and decidability of Tarskian

2.2. STRONG DEPTH-BOUNDED APPROXIMATIONS

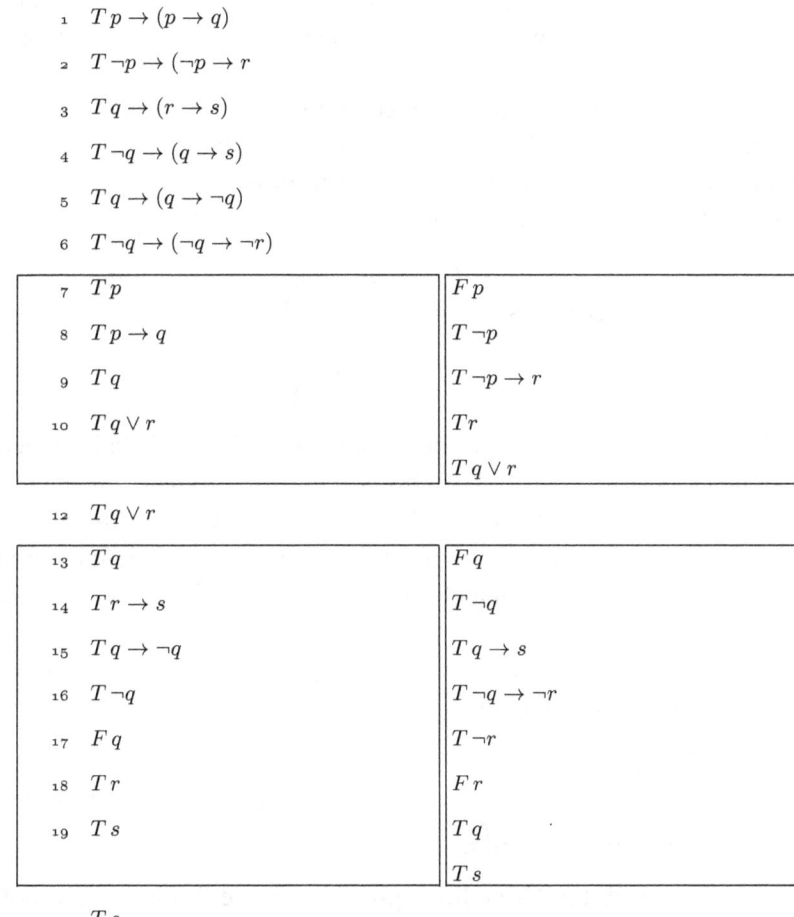

Figure 2.8: A normal non-analytic C-intelim$^+$ proof that cannot be transformed into an analytic one without increasing its depth.

strong depth-bounded approximations, suggesting that, while being worthy of logical investigation for their own sake, they may not be suitable to define approximation systems in our sense. In fact, it is open whether a general canonical search space can be specified, so that a further generalized version of the SFP can be proven with no need for bounding the search space *a priori*.

We conjecture that some more general notion of canonical search space can be defined for $^+\vdash_k^g$ so that the size of the search space can be polynomially bounded without loss of deductive power, in which case each Tarskian depth-bounded consequence relation would indeed be tractable:

Conjecture 2.2.7. *For every polynomially bounded $g \in \mathcal{V}$ there exists a polynomially bounded $f \in \mathcal{S}^g$ such that*

$$^+\vdash_k^{F,g} = {^+\vdash_k^{f,g}},$$

where F is the function that yields the set of all formulae of $\mathcal{L}$. Independently of how the conjecture is settled, it makes sense to introduce a deducibility relation $^+\vdash_k^{f,g}$ that allows us to specify the search space function as well as the virtual space function, so as to match the relation $^+\models_k^{f,g}$ defined in Section 2.2.2. A suitable choice may be setting it so as to deliver the canonical search space of (2.11).

We now update our notion of k-depth hyper-proof to accommodate the specification of the search space.

Definition 2.2.8. *A k-depth C-intelim$^+$ proof of ψ from X of type $\langle f, g \rangle$ is a k-depth C-intelim$^+$ proof of φ from X of type g, such that all the S-formulae occurring in it are in $f(X^u \cup \{\varphi^u\})$.*

If $f = g = \text{sub}$ the C-intelim$^+$ proof in Figure 2.8 is not a proof of type $\langle f, g \rangle$, because $q \vee r$ is *not* a subformula of the premises or of the conclusion. On the other hand, if we expand the search space, for instance using the function sub_1 (see p. 72), then $q \vee r$ does belong to the search space.

The proof of soundness and completeness of $\vdash_k^{f,g}$ with respect to $^+\models_k^{f,g}$ is routine:

Proposition 2.2.9. $\Gamma\, ^+\vdash_k^{f,g} A$ *if and only if* $\Gamma\, ^+\models_k^{f,g} A$.

2.2. STRONG DEPTH-BOUNDED APPROXIMATIONS

When $f = g = \text{sub}$, a decision procedure for $\Gamma \; {}^+\vdash_k^{f,g} A$ can be obtained from that for $\Gamma \vdash_0 A$ discussed in Section A.3 by adapting the Expand sub-routine of the algorithm (see Section A.6 in the Appendix). It can be shown that:

Proposition 2.2.10. *If $f = g = \text{sub}$, the complexity of the decision of $\Gamma \; {}^+\vdash_k^{f,g} A$ is $O(n^{2k+2})$, where $n = |\Gamma \cup \{A\}|$*

Thus, the hierarchy of the ${}^+\vdash_k^{\text{sub},\text{sub}}$ relations provides an approximation system to full Boolean Logic whose semantic counterpart is given by the strong depth-bounded relations ${}^+\models_k^{\text{sub},\text{sub}}$. When $g \not\trianglelefteq \text{sub}$, the full (yet polynomially bounded) search space yielded by the canonical search space function f^g (2.11) should be visited in the decision procedure and n will be equal to its total size.

C-intelim${}^+$ proofs of type $\langle f, g \rangle$ can also provide a Boolean basis for the proof theory of logics that do no have the subformula property with no need for ingenious, albeit often contrived and *ad hoc*, presentations devised for the only purpose of satisfying this property.

Algorithm A.6.1 can be adapted to show tractability whenever the size of the search space (and therefore of the virtual space) is bounded above by a polynomial in the size of the input set.

Proposition 2.2.11. *For every $f \in S^g$ and $g \in \mathcal{V}$ such that g is polynomially bounded, the relation ${}^+\vdash_k^{f,g}$ is tractable.*

Conjecture 2.2.7, if true, would imply, via Proposition 2.2.11 that each ${}^+\vdash_k^g$ is a tractable Tarskian consequence relation, with no need for bounding the search space *ad hoc*. Albeit, theoretically interesting, for most practical applications, in which there is a natural way of bounding the search space for the problems of concern, settling the conjecture is not crucial.

C-intelim tableaux (or C-intelim${}^+$ proofs) are natural deduction systems based on a set of inference rules that is more suitable for classical propositional logic and on a single, structural, "discharge" rule, namely the RB rule. Such proofs can be easily transformed into the more conventional format of natural deduction in the style of Gentzen and Prawitz where proofs are trees that, unlike C-intelim tableaux, grow "upward", the root contains the conclusion and the premises are in the leaves. In (D'Agostino *et al.*, 2020) we provide a detailed proof-theoretical analysis of proof-trees in this format based on the C-intelim rules.

Let us briefly review the main points of this analysis. First, what is the exact relation between C-intelim hyper-sequences and C-intelim tableaux? We know that the former are deductively more powerful than the latter, but do not seem to enjoy the generalized subformula property, which prompted us to introduce an explicit restriction on the search space. However, it is not difficult to see that C-intelim tableaux are nothing but (a notational variant) of a *special kind* of C-intelim$^+$ proofs.

Let us say that a C-intelim$^+$ proof Π of φ from X is *quasi-normal* if and only if it satisfies the following conditions:

- no application of RB in Π is followed by an application of an inference rule;

- for no applications of XFQ in Π its conclusion is $T \curlywedge$, and for no application of XFQ its conclusion is a premise of an application of an inference rule;

- Π is strongly non-redundant.

Moreover, we say that π is *g-normal* if every application of RB in π is g-analytic. When $g = \mathsf{sub}$ we just say that it is normal.

For example, the C-intelim$^+$ proof in Figure 2.7 is normal and it can be immediately seen that it is just a C-intelim tableau, except for the inessential use of the $\curlywedge$-rules. Hence, as a consequence of Proposition 2.1.19, g-normal C-intelim$^+$ proofs do have the GSFP and the search space can be safely assumed to be the canonical one in (2.11), with no loss of deductive power.

2.3 Non-contamination

The restriction of C-*intelim* to normal proofs enforces a stricter control discipline on proof-construction, to the effect that such proofs, besides enjoying the SFP, also enjoy a kind of weak relevance property that we call *non-contamination*.

2.3. NON-CONTAMINATION

2.3.1 The contamination problem.

Let us say that two formulae are *syntactically disjoint* if they share no atomic formula. Two *sets* Γ and Δ are *syntactically disjoint* if every formula of Γ is syntactically disjoint from every formula of Δ. In what follows we shall write "$\Gamma \parallel \Delta$" for "Γ is syntactically disjoint from Δ" and denote by $\vdash_C$ the consequence relation of classical propositional logic.

It is routine to show that the following holds by classical semantics:

Proposition 2.3.1. *For every Γ and Δ, if*

1. $\Delta \parallel \Gamma \cup \{A\}$, *and*

2. $\Gamma \cup \Delta \vdash_C A$,

then at least one of the following holds true:

- Γ *is consistent and* $\Gamma \vdash_C A$;

- Γ *is inconsistent and* $\Gamma \vdash_C A$;

- Δ *is inconsistent and* $\Delta \vdash_C A$.

This holds for every (possibly empty) Γ, Δ. The special case in which Δ is empty is trivial. In the special case in which Γ is empty, the proposition implies that:

Corollary 2.3.2. *If $\Delta \parallel \{A\}$, then $\Delta \vdash_C A$ if and only if either A is a tautology or Δ is inconsistent.*

The *contamination problem* is the problem that arises in a proof system S when:

1. for some *non-empty* Δ such that $\Delta \parallel \Gamma \cup \{A\}$, with $A \neq \curlywedge$, we have an S-proof of A depending[16] on $\Gamma \cup \Delta$, or

2. for some *non-empty* Δ, Γ such that $\Delta \parallel \Gamma$, we have an S-proof of $\curlywedge$ depending on $\Gamma \cup \Delta$.

---

[16]By this we mean that all the assumptions are actually used in the deduction tree.

In case 1, it may be that $\Gamma \vdash_C A$, in which case the unrelated assumptions in Δ are clearly unnecessary to obtain the conclusion. Or it may be that $\Gamma \nvdash_C A$, in which case, by classical semantics, Δ is inconsistent and assumptions that are totally unrelated either to the conclusion or to Γ play an active role in the proof to obtain a conclusion that could not have been otherwise obtained from Γ. The paradigmatic example is:

(2.13)
$$\frac{\dfrac{A \quad \neg A}{\curlywedge}}{B.}$$

Here the conclusion is obtained from the premises by means of a practically meaningless use of *ex-falso quodlibet*.[17]

As Michael Dummett once put it:

> Obviously, once a contradiction has been discovered, no one is going to go *through it*: to exploit it to show that the train leaves at 11:52 or that the next Pope will be a woman (Dummett, 1991b, p. 209).

In case 2, either Δ or Γ must be inconsistent on their own and, again, unrelated and redundant assumptions are used to obtain the proof of $\curlywedge$. In any case such a proof violates a basic relevance condition that can indeed be satisfied by a better-behaved proof, except for one distinguished case in which all of the following hold true:

1. Γ is empty,

2. Δ is inconsistent and cannot be partitioned into two syntactically disjoint subsets,

3. $A \neq \curlywedge$ and is obtained from Δ by means of an *ex-falso* (or better *ex-contradictione*) inference.

---

[17] A controversial justification of (2.13) is provided by the celebrated "Lewis proof" (p. 48 above). See Bennett (1969) for a thorough discussion in which the author provides an interesting example which does not appear to be counterintuitive argues that "the logic of the Lewis argument can be displayed in indefinitely many other examples, whose validity is highlighted by the their being possible — even plausible — slices of real argumentative life" (p. 220).

2.3. NON-CONTAMINATION

This kind of "proofs" are, in essence, the ones that we have called *improper* in the context of C-intelim (p. 77). The simplest example is the inference shown in (2.13).

While such inferences cannot be expunged from any *complete* system for classical logic, we can easily obtain from them a proof that Δ is inconsistent, that is a proof of $\curlywedge$ depending on Δ.

Definition 2.3.3 (Contaminated proofs). *Given a proof system S, we say that an S-proof π of A depending on Γ is* contaminated *if one of the following two conditions hold:*

1. *$A \neq \curlywedge$ and for some non-empty $\Delta \subseteq \Gamma$, $\Delta \parallel (\Gamma \setminus \Delta) \cup \{A\}$;*

2. *$A = \curlywedge$ and for some non-empty $\Delta \subset \Gamma$, $\Delta \parallel (\Gamma \setminus \Delta)$.*

In both cases we call Δ a contaminating set *for π.*

Definition 2.3.4 (R-contaminated proofs). *We say that a proof π is* redundantly contaminated *(R-contaminated for short) if (i) π is contaminated and (ii) there is a contaminating set Δ for π such that $\Delta \subset \Gamma$.*

Note that our definition implies that any contaminated *refutation* of Γ (a proof of $\curlywedge$ depending on Γ) is always R-contaminated. Proofs that are contaminated but not R-contaminated correspond to the exceptions discussed above that cannot be expunged by any proof system that is complete for classical logic (or for the classical depth-bounded approximations which are still explosive). They can indeed be expunged if we switch to a paraconsistent system, such as the one that results from restricting C-intelim or C-intelim[+] to *proper* normal proofs or from a classical extension of the system discussed in Tennant (1987).

It turns out that the restriction of C-intelim to normal proofs[18] delivers only proofs that are not R-contaminated. By contrast, the notion of normal natural deduction proof Prawitz (1965) is not sufficient for this purpose. The trees shown in Figure 2.9 represent R-contaminated proofs that are normal in Prawitz's sense (assuming that B and E are atomic to satisfy the restriction on the $\curlywedge_C$ rule, see Prawitz (1965, Chapter III)). Note that such proofs

---

[18]In this section for ease of exposition we shall switch to unsigned formulae and assume the reader to be able to adapt all the previous definitions.

$$\dfrac{A \to B \qquad \dfrac{\dfrac{C \quad \neg C}{\curlywedge}}{A}}{B} \qquad\qquad \dfrac{\dfrac{\dfrac{C \quad \neg C}{\curlywedge}}{B} \qquad \dfrac{\dfrac{\dfrac{D \quad \neg D}{\curlywedge}}{E} \qquad E \to \neg B}{\neg B}}{\curlywedge}$$

Figure 2.9: R-contaminated normal proofs in Prawitz's natural deduction.

can be turned into proofs of $\curlywedge$ depending on the *same* set of assumptions. On the other hand, (D'Agostino *et al.*, 2020) shows that normal C-intelim proofs are never R-contaminated. Let us focus on the "exceptional" proofs that are contaminated but not R-contaminated. These are proofs of A depending on a non-empty $\Gamma \parallel \{A\}$ in which $A \neq \curlywedge$ and Γ cannot be partitioned into syntactically disjoint subsets. For a *normal* C-intelim proof this situation obtains only when the proof is improper and therefore is essentially a proof of $\curlywedge$ depending on the same set Γ of assumptions.

We have already commented that while such proofs express classically valid inferences, they are devoid of any practical value, in that they cannot be used to make any reasonable inference at all.[19]

Definition 2.3.5. *A* C-intelim tableau *proof is* proper *if at least one of its branches is open. The corresponding normal C-intelim$^+$ proof is proper if at least one of the auxiliary proofs in its innermost boxes does not end with an application of* XFQ*,*

For example, the proof in Figure 2.7 is proper. Observe that, in normal C-intelim$^+$ proofs, XFQ can be applied only as the last step of one of its innermost boxes. Observe also that every normal refutation of Γ (i.e., a proof of $\curlywedge$ depending on Γ) is proper.

Remark 2.3.6. *Observe that:*

[19] This view is shared, of course, by all advocates of relevance logic. However, with the notable exceptions of Timothy Smiley and Neil Tennant, most of them argue that disjunctive syllogism should be rejected as well as the ex-falso rule, given the role played by the latter in the proof of the former within the framework of Gentzen-style natural deduction. By contrast, disjunctive syllogism is a primitive rule in the C-*intelim* system.

2.3. NON-CONTAMINATION

- *Improper C-intelim tableau proofs of A from Γ are sic et simpliciter normal refutations of Γ.*

- *Every improper normal C-intelim$^+$ proof of A from Γ can be transformed into a normal C-intelim$^+$ proof of $\curlywedge$ from Γ with at most a linear overhead (just remove all the final applications of* XFQ *in the innermost boxes, and replace all the occurrences of A as conclusion of an RB applications with $\curlywedge$).*

2.3.2 Variable sharing and non-contamination

The results in (D'Agostino *et al.*, 2020) can be easily adapted to show that normal C-*intelim* proofs are never R-contaminated and that *proper* normal proofs are never contaminated, so that they enjoy the variable-sharing property, except for the case in which their conclusion is $\curlywedge$. More specifically, for every depth-bounded approximation it holds that:

Proposition 2.3.7. *For every non-empty Γ and every A, there is no proper normal proof of A depending on Γ such that $A \neq \curlywedge$ and $\Gamma \parallel \{A\}$.*

It follows that in a normal k-depth proof from a non-empty Γ, if $\Gamma \parallel \{A\}$, either $A = \curlywedge$ or the proof is improper, that is, the proof ends with an application of XFQ. Hence, Γ is inconsistent.

Corollary 2.3.8. *For every non-empty Γ and every A, if there is a normal k-depth proof of A depending on Γ such that $\Gamma \parallel \{A\}$, then Γ is inconsistent.*

Corollary 2.3.9 (Variable-sharing property). *If there is a proper normal k-depth proof of $A \neq \curlywedge$ from Γ, then $\Gamma \nparallel \{A\}$.*

Although proper normal proofs are not complete for classical logic, nor are they for for its depth-bounded approximations, they are complete for the valid inferences from consistent sets of assumptions, since if there is no proper normal proof of A from Γ, then Γ must be inconsistent, in that any improper proof, by Remark 2.3.6, "contains" a refutation of Γ. As stated above, they also enjoy the variable sharing property. The system of deduction that accepts as admissible only proper normal proofs has close connections with Tennant-style relevance logic (Tennant, 1984, 1987). These connections will be investigated in future work.

It follows from Definitions 2.3.3 and 2.3.4 that:

Proposition 2.3.10 (Non-contamination property)**.** *In the C-intelim system, for every k-depth subsystem:*

- *No normal proof of A from Γ is R-contaminated.*
- *No proper normal proof of A from Γ is contaminated.*

Part II

Applications of depth-bounded reasoning in artificial intelligence and philosophy

Chapter 3

Rational non-monotonic reasoning

3.1 Introduction

In this chapter we look to applications of the previous chapters' account of proof systems for depth-bounded deduction. In particular, we show how Chapter 2's C-*intelim*, more specifically its unsigned version using the rules in Table 1.6), restricted to normal proofs that are non-contaminated (Section 2.3), can be deployed for reasoning by resource bounded agents, *without compromising on the rationality of the outcome of reasoning*. In this chapter, since we will not deal with S-formulae and for the sake of uniformity with previous papers[1] we shall use lower case greek letters as metalinguistic variables ranging over $\mathcal{L}$-formulae.

The *GOFAI* ("Good Old Fashioned AI") programme conceived of agents deploying logical formalisms that accommodate updates to their symbolic model of the world (their beliefs) on the basis of perceived information. Reasoning with these beliefs in the context of desires (motivational states) adopted as goals, agents commit to intentions (high level symbolic plans) that initiate action and thus changes to the world and further sensory updates; the sense-reason-act cycle continues. The deductive, and in particular classical logic paradigm was clearly not suitable for such "real-world"

---

[1] For example (D'Agostino & Modgil, 2016, 2018a).

agents operating on the basis of incomplete and possibly erroneous information about the world. Inferences are defeasible; they need to be withdrawn as further information is acquired so as to revise or augment an agent's beliefs, either because these beliefs are in error or incomplete, or because the world itself changes. The monotonicity of classical logic, whilst unproblematic in the context of axiomatic descriptions of idealised and unchanging models, does not allow for the withdrawing of previously held conclusions. Hence, since the late 70s and early 80s, logicians, philosophers and computer scientists have developed non-monotonic logics, primarily as a sub-area of symbolic AI.

Broadly speaking two approaches have gained popularity.[2] The "maxi-consistent" approach (e.g., *Preferred Subtheories*, Brewka 1989) uniformly represents an agent's "belief base" as a finite set of possibly inconsistent classical logic formulae $\mathcal{B}$, all of which are effectively interpreted as defeasible. One then selects, from amongst the maximally consistent subsets (*mcs*) of $\mathcal{B}$, those — $E_1, \ldots, E_n$ — that are maximal under some preference ordering over the *mcs*. This ordering may in turn be lifted from a priority ordering over the individual formulae that encodes the relative entrenchment of beliefs, or that respects some conflict resolution principle such as the temporal (beliefs acquired later take precedence over earlier conflicting counterparts) or specificity principle (elucidated below). One then identifies alternative sets of credulous (*cr*) non-monotonic (*nm*) inferences, where each such set equates with the closure (under classical consequence) of each E_i (i.e., $Cn(E_i)$).

$\mathcal{B} \mathrel{\mid\!\sim}_{cr} \alpha$ iff $\exists E_i, \alpha \in Cn(E_i)$
(*credulous nm consequence relation*).

The sceptical (*sc*) *nm* inferences are those belonging to $\bigcap_{i=1}^{n} Cn(E_i)$

$\mathcal{B} \mathrel{\mid\!\sim}_{sc} \alpha$ iff $\forall E_i, \alpha \in Cn(E_i)$
(*sceptical nm consequence relation*)

and is arguably more useful from the perspective of an agent needing to commit to a particular world view.

---

[2] Aside from "model theoretic approaches", such as circumscription, (McCarthy, 1986), that effectively identify the preferred — as in most typical — models that correspond to the given set of formulae (cf. preferential model semantics for non-monotonic logics, Kraus *et al.* (1990)).

3.1. INTRODUCTION

For example, given the belief base:

$$\{\forall x.p(x) \rightarrow b(x), p(tweety), \forall x.b(x) \rightarrow f(x), \forall x.p(x) \rightarrow \neg f(x)\}$$

where p, b and f are, respectively, shorthand for predicates *penguin*, *bird* and *fly*. Then the two sets of credulous *nm* inferences are

$$E_1 = Cn(p(tweety), \forall x.p(x) \rightarrow b(x), \forall x.b(x) \rightarrow f(x)) \text{ and}$$
$$E_2 = Cn(p(tweety), \forall x.p(x) \rightarrow b(x), \forall x.p(x) \rightarrow \neg f(x)).$$

Neither $f(tweety)$ or $\neg f(tweety)$ are sceptical *nm* inferences. If $\forall x.p(x) \rightarrow \neg f(x)$ is prioritised above $\forall x.b(x) \rightarrow f(x)$, based on the specificity principle giving precedence to properties of sub-classes over properties of their super-classes, then the single set of credulous and sceptical *nm* inferences is E_2 (penguins are a special sub-class of birds that do not fly).

The second "defeasible inference rule" approach effectively models a belief base in terms of a consistent "world description" W, typically encoded as a set of classical formulae, together with additional domain specific *inference rules* for drawing defeasible inferences (e.g., $\frac{bird(X)}{fly(X)}$), contingent on checking that such inferences are not in conflict with (i.e., are consistent with) W. Well known examples of this approach are the various flavours of Reiter's *Default Logic*, (Reiter, 1980), and the closely related *Autoepistemic Logic*, (Marek & Truszczyski, 1991). Additionally, one needs to resolve conflicts between contradictory defeasible inferences, based on the kinds of prioritisation principles alluded to above. One thus obtains possibly multiple extensions — $E_1, \ldots, E_n$ — of classical formulae that similarly yield credulous and sceptical *nm* inferences. Each E_i can be interpreted as a possible world model[3] represented by a deductively closed maximal consistent subset of defeasible inferences, each consistent with W.

However, the last two decades have witnessed a waning of interest in non-monotonic logics for agent reasoning. This is arguably in large part due to two reasons. Firstly, there is the unacceptable computational demand exacted when reasoning non-monotonically over realistically complex and rich belief bases, due to the assumption motivating the study of depth-bounded

---

[3]Not in the Kripke-ian modal semantics sense, but rather as formalised in preferential model semantics for non-monotonic logics, (Kraus *et al.*, 1990). However modal semantics have been posited for non-monotonic logics (e.g., McDermott 1982).

logics in previous chapters — that a rational agent's beliefs are deductively closed (the assumption of *logical omniscience*) — and that is further exacerbated by the demands imposed by the consistency checking required to partition the belief bases into the maximally consistent $E_1, \ldots, E_n$ in maxiconsistent approaches, or checking consistency of defeasible inferences with W in defeasible rule approaches.

Secondly, the increasing dominance and success of machine learning (in particular deep learning) applications has reinforced scepticism as to whether the symbolic AI paradigm — and more specifically non-monotonic logics — represent the most promising approach to implementation of single agent reasoning. This is evidenced not only by the media hype around AI, which almost exclusively points to eye-catching success stories based predominantly on neural network implementations (e.g., Google Deep Mind's AlphaFold https://alphafold.ebi.ac.uk/ and the astonishing impact of large language models) but also the dominance of machine learning in major AI conferences (accompanied by a decline in publications on logic-based approaches).

3.2 Argumentation, non-monotonic logic and dialogue

It is in this context that we leverage the use of depth-bounded logics to advance a programme of research aiming at revitalising the role of non-monotonic logics in AI. While it is at least conceivable that such logics may play a more limited role in single agent reasoning than that originally conceived by the GOFAI programme (although many convincingly argue that integration of symbolic and machine learning systems will be required for complex forms of agent reasoning (Besold *et al.*, 2017; Marcus, 2018), it is arguably far less contentious to claim that symbolic, and in particular non-monotonic logics, will be required to support joint reasoning amongst human and artificial agents. After all, more complex reasoning tasks that involve managing uncertainty and resolving conflicts, often benefit from input and insights elicited from multiple human agents engaged in the dialogical exchange of natural language locutions. Hence, non-monotonic logics can provide normative guidance for rational joint deliberation amongst human agents, and amongst human and artificial agents, where the latter will *necessarily* need to communicate symbolically, in ways understandable to their

3.2. ARGUMENTATION AND NON-MONOTONIC LOGIC

human interlocutors (Modgil, 2017a). Indeed, the latter may be of particular importance if AI reasoning and decision making is to be aligned with human values (Modgil, 2017a, 2018; Bezou-Vrakatseli *et al.*, 2024).[4] Ensuring that such dialogical exchanges comply with rational principles governing the handling of uncertainty and resolution of conflicts, requires development of formal models of distributed non-monotonic reasoning that can serve to constrain and guide the choice of locutions and their relationships in such dialogues. It is arguably this requirement that is best served by argumentation-based characterisations of non-monotonic logics.

3.2.1 Argumentative characterisations of NML

Dung's seminal theory of argumentation (Dung, 1995) provides for characterisations of both the defeasible inference rules and maxiconsistent approaches. Non-monotonic reasoning is fundamentally concerned with the recognition and resolution of conflict or inconsistency; the withdrawing of an inference α is instigated by the recognition that some inference β is (given what one commits to as certain or indefeasible) inconsistent with α, and then arbitrating in favour of β. The recognition and resolution of conflicts is also central to the enterprise of argument and debate, and it is this insight that Dung's theory formally realises. For example, consider some maxiconsistent non-monotonic consequence relation $\mathcal{B} \mathrel{|\!\sim} \alpha$. Now, classical logic based arguments (Amgoud & Cayrol, 2002; Gorogiannis & Hunter, 2011; Modgil & Prakken, 2013) can be defined on the basis of $\mathcal{B}$:

---

[4]Cooperative Reinforcement Learning (Hadfield-Menell *et al.*, 2016) is amongst the most prominent approaches to ensuring that the values of machines are aligned with human values. The idea is that the primary goal of AI systems should be to maximise the human reward function while always remaining uncertain as what the humans' preferences/values are. This in turn incentivises ongoing learning of preferences and values through interaction with humans. The challenge is to then ensure that this learning imperative not impede AI systems from simultaneously performing the tasks for which they were designed. Now, if AI systems supporting decision making were to arrive at decisions, via collaborative dialogues that include eliciting human input relevant to the preferences/values pertinent to decisions, then preferences/values are learnt while decision making tasks are simultaneously performed. Hence the requirement for dialogical exchanges supporting distributed non-monotonic reasoning.

(3.1) $$Args = \{(\Delta, \phi)| \; 1)\; \Delta \subseteq \mathcal{B},\; 2)\; \Delta \vdash \phi,\; 3)\; \Delta \not\vdash \bot \text{ and}$$
$$\neg \exists \Delta' \subset \Delta \text{ s.t. } \Delta' \vdash \phi\}.$$

That is, an argument X consists of 1) premises $\text{prem}(X) = \Delta$ that are a subset of a belief base $\mathcal{B}$, and that 2) classically entail the argument's conclusion $\text{conc}(X) = \phi$, and 3) is a valid argument only if $\text{prem}(X)$ is classically consistent (*Prem-con*), and no proper subset of $\text{prem}(X)$ entail ϕ so that all premises in $\text{prem}(X)$ can be said to be relevant to concluding $\text{conc}(X)$ (*Prem-rel*).

For any $X, Y \in Args$, X is said to *attack* (i.e., "challenge" or represent a "counter-argument to") Y iff $\text{conc}(X) = -\alpha$, $\alpha \in \text{prem}(Y)$, where $-\alpha = \beta$ if α is of the form $\neg \beta$, else $-\alpha = \neg \alpha$. X is then said to attack Y on $Y' = (\{\alpha\}, \alpha)$, where Y' is the "elementary" argument constituted by the targeted premise. The success of an attack may be contingent on a preference relation over arguments, which itself may be lifted from a priority ordering $\leq$ over the formulas in $\mathcal{B}$ (where as usual, $\alpha < \beta$ iff $\alpha \leq \beta$ and $\beta \not\leq \alpha$), so that X successfully attacks (*defeats*) Y iff $X \not\prec Y'$.

(3.2) $\forall X, Y \in Args$: X defeats Y iff X attacks Y on Y' and $X \not\prec Y'$.

An example is a preference relation given by the so called *Elitist* lifting (Modgil & Prakken, 2013):

(3.3) $\quad X \prec Y$ if $\exists \alpha \in \text{prem}(X), \forall \beta \in \text{prem}(Y), \alpha < \beta$.

Intuitively, Y is strictly preferred to X if a premise in X is strictly dominated by (i.e., $<$) all premises in Y.

A Dung *Argumentation Framework* (AF) defined by $\mathcal{B}$ and a preference relation over the set $Args$ defined by $\mathcal{B}$, is a directed graph ($Args, Def$) where $Def = \{(X, Y)|X \text{ defeats } Y\}$ is the binary defeat relation over $Args$. One can then determine the *admissible* sets of arguments $E \subseteq Args$, each of which must be *conflict free* — $\neg \exists X, Y \in E$ such that $(X, Y) \in Def$ — and each of which defends its members against all defeats (in which case its members are said to be acceptable w.r.t. E):

(3.4)
$E \subseteq Args$ is admissible iff E is conflict free and $\forall Y \in E, \forall X \in Args$:
$$X \text{ defeats } Y \text{ implies } \exists Z \in E, Z \text{ defeats } Y.$$

3.2. ARGUMENTATION AND NON-MONOTONIC LOGIC 113

Given this core intuitive notion of conflict free sets of acceptable, or defendable, sets of arguments (referred to as "admissible extensions") one can further identify the notion of a *complete* extension E as an admissible extension E that maximally includes arguments that it defends (i.e., E includes all arguments acceptable w.r.t. E):

> An admissible extension $E \subseteq Args$ is a complete extension iff $\forall X$ s.t. X is acceptable w.r.t. E, $X \in E$.

For example, given an AF $(\{A, B, C\}, \{(A, B), (B, C)\})$ (see Figure 3.1a), then $\{A\}$ is admissible but not complete, and $\{A, C\}$ is both admissible and complete.

In addition to what Dung refers to as extensions defined under the admissible or complete "semantics", complete extensions can be further distinguished according to whether they are maximal, or minimal, under set inclusion, and whether all arguments that are not in a complete extension E are defeated by an argument in E:

> A complete extension E is: 1) an extension under the *preferred* semantics if $\neg \exists E' \supset E$ s.t. E' is complete; 2) an extension under the *grounded* semantics if $\neg \exists E' \subset E$ s.t. E' is complete; 3) an extension under the *stable* semantics if $\forall Y \notin E, \exists X \in E$, s.t. X defeats Y.

Notice that it can be shown that there is only ever one grounded extension of an AF; intuitively, the grounded semantics are inherently sceptical, whereas the preferred/stable are credulous. For example, consider the AF in Figure 3.1b. There are two preferred and stable extensions – $\{A, D\}$ and $\{B, D\}$ — and the grounded extension is the empty set (no arguments are acceptable w.r.t. $\emptyset$ which is also, trivially, conflict free and admissible).

The argumentation-based inferences from $\mathcal{B}$ are then defined, under the various semantics, credulously or sceptically. Let $E_1, \ldots, E_n$ be the s-extensions of an $AF = (Args, Def)$ whose arguments and defeats are defined by $\mathcal{B}$ and a preference relation over the set $Args$, and where $s \in \{grounded, preferred, stable\}$. Let cr and sc respectively denote "credulous" and "sceptical":

(3.5) $\qquad \mathcal{B} \mathrel{\vert\!\sim}^s_{cr} \phi$ iff $\exists$ an s-extension $E_i, \exists X \in E_i, \text{conc}(X) = \phi$

114   CHAPTER 3. RATIONAL NON-MONOTONIC REASONING

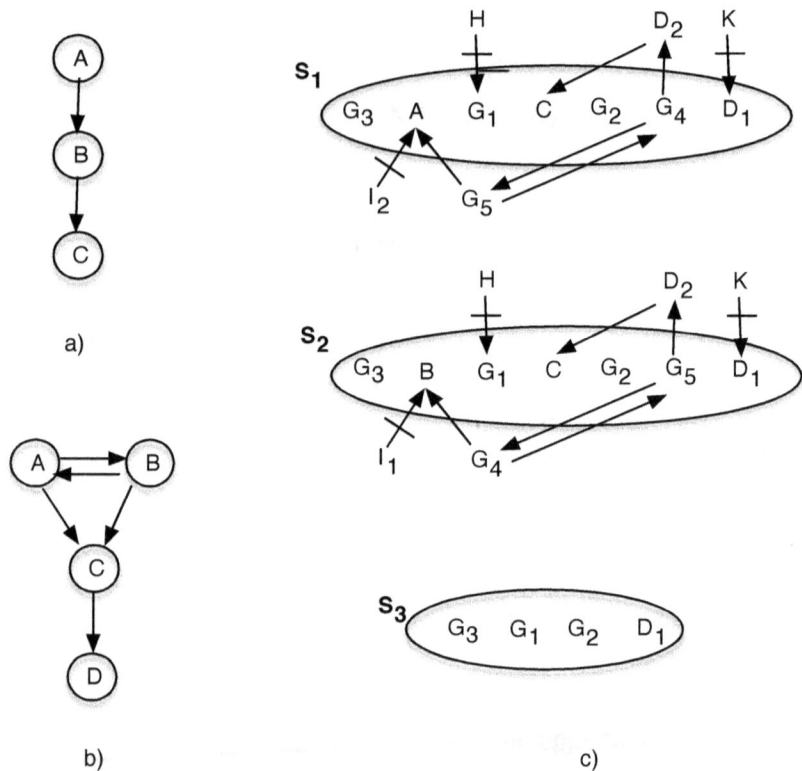

Figure 3.1: Arrows denote defeats. Bisected arrows denote attacks that fail to succeed as defeats. In c), S_1 and S_2 are subsets of the two preferred/stable extensions of the argument framework instantiated by the ordered belief base in Example 3.2.1, where the arguments are defined (see Eq. 3.1) as in standard approaches to classical logic argumentation (Amgoud & Cayrol, 2002; Gorogiannis & Hunter, 2011; Modgil & Prakken, 2013). S_3 is a subset of the grounded extension.

3.2. ARGUMENTATION AND NON-MONOTONIC LOGIC

(3.6) $\quad \mathcal{B} \mid\!\sim^{s}_{sc} \phi$ iff $\forall$ s-extensions $E_i, \exists X \in E_i, \mathrm{conc}(X) = \phi$

One can then show correspondence results for argumentative characterisations of both the defeasible inference rule and maxiconsistent non-monotonic logics.[5] For example, *Preferred Subtheories* (*PS*) is a maxiconsistent approach that exploits a total ordering over a classical logic belief base $\mathcal{B}$, so as to identify the preferred maximal consistent subsets (*mcs*) of $\mathcal{B}$, and the credulous and sceptical consequence relations ($\mid\!\sim^{PS}_{cr}$ and $\mid\!\sim^{PS}_{sc}$) as described in Section 3.1 (Brewka, 1989).[6] It can be shown that for the argumentation based inferences defined by $\mathcal{B}$ (where the preference relation over arguments is defined by the Elitist lifting of the total ordering over $\mathcal{B}$):

(3.7) $\quad \mathcal{B} \mid\!\sim^{PS}_{cr/sc} \phi$ iff $\mathcal{B} \mid\!\sim^{s}_{cr/sc} \phi$ ($s \in \{preferred, stable\}$)[7]

Example 3.2.1. *Let $\mathcal{B}$ =*

$$\{a, b, c, (a \wedge b) \rightarrow \neg c, g, g \rightarrow \neg a \vee \neg b\}$$

where $\leq$ is a total ordering over $\mathcal{B}$ such that ($\alpha \approx \beta$ iff $\alpha \leq \beta$ and $\beta \leq \alpha$):

$$c < a \approx b \approx (a \wedge b) \rightarrow \neg c < g \approx g \rightarrow \neg a \vee \neg b$$

and a, b, c are propositional atoms respectively denoting "attend conference A,B,C" and g denotes "the conference budget = €2000". The ordering over $\mathcal{B}$ induces two preferred mcs *("preferred subtheories")*

$$E_1 = \{c, a, (a \wedge b) \rightarrow \neg c, g, g \rightarrow \neg a \vee \neg b\}$$

---

[5] E.g., Preferred Subtheories (Brewka, 1989) and Prioritised Default Logic (Brewka, 1994) as shown in (D'Agostino & Modgil, 2018b) and (Young *et al.*, 2016) respectively.

[6] Preferred Subtheories is amongst the most influential approaches to non-monotonic reasoning, as well as belief revision, and has also been used for reasoning about time, reasoning by analogy, reasoning with compactly represented preferences, judgment aggregation, and voting (Lang, 2015).

[7] For s = stable, the correspondence is shown in (Modgil & Prakken, 2013; Thang & Luong, 2014). For $s = preferred$ the correspondence is shown in (D'Agostino & Modgil, 2018b) (i.e., D'Agostino & Modgil 2018b show that the stable and preferred extension coincide).

and
$$E_2 = \{c, b, (a \wedge b) \to \neg c, g, g \to \neg a \vee \neg b\}$$
and so
$$\mathcal{B} \mid\!\sim_{cr}^{PS} \phi, \phi \in Cn(E_1) \cup Cn(E_2)$$
whereas
$$\mathcal{B} \mid\!\sim_{sc}^{PS} \phi \in Cn(\{c, (a \wedge b) \to \neg c, g, g \to \neg a \vee \neg b\}), \mathcal{B} \not\mid\!\sim_{sc}^{PS} \phi \in Cn(\{a, b\})$$

Intuitively, the budget of €2000 is a hard constraint precluding attendance at both conferences A and B, and given that attendance at A and B precludes attendance at C, then despite attendance at A and attendance at B being preferred to attendance at C, one can alternatively attend A and C or B and C.

The arguments Args defined by $\mathcal{B}$ are the tuples (Δ, ϕ), $\Delta \in 2^{\mathcal{B}}$, and include (but are of course not limited to):

$A = (\{a\}, a)$,
$B = (\{b\}, b)$,
$C = (\{c\}, c)$,
$G_1 = (\{g\}, g)$,
$G_2 = (\{g \to \neg a \vee \neg b\}, g \to \neg a \vee \neg b)$,
$D_1 = (\{(a \wedge b) \to \neg c\}, (a \wedge b) \to \neg c)$
and
$G_3 = (\{g, g \to \neg a \vee \neg b\}, \neg a \vee \neg b)$,
$G_4 = (\{g, a, g \to \neg a \vee \neg b\}, \neg b)$,
$G_5 = (\{g, b, g \to \neg a \vee \neg b\}, \neg a)$,
$D_2 = (\{a, b, (a \wedge b) \to \neg c\}, \neg c)$,
$H = (\{a, b, g \to \neg a \vee \neg b\}, \neg g)$,
$I_1 = (\{c, a, (a \wedge b) \to \neg c\}, \neg b)$,
$I_2 = (\{c, b, (a \wedge b) \to \neg c\}, \neg a)$,
$K = (\{c, a, b\}, \neg((a \wedge b) \to \neg c))$.

Figure 3.1c shows subsets S_1 and S_2 of the preferred and stable extensions, and S_3 a subset of the grounded extension. Notice that:

$$\{(H, G_1), (I_2, A), (I_1, B), (K, D_1)\} \not\subseteq Def$$

3.2. ARGUMENTATION AND NON-MONOTONIC LOGIC

since $a, b < g$ and $c < a$ and $c < b$ and $c < (a \wedge b) \rightarrow \neg c$ so that (according to the Elitist lifting — see Eq. 3.3) $H \prec G_1, I_2 \prec A, I_1 \prec B$ and $K \prec D_1$ respectively.

Observe that the correspondence result in Eq. 3.7 holds for $s \in \{stable, preferred\}$. However, only the right to left half of the correspondence holds for the grounded semantics since in the example above $\mathcal{B} \hspace{1pt}\vert\!\sim^{PS}_{sc} c$ but $\mathcal{B} \hspace{1pt}\not\vert\!\sim^{grounded}_{sc} c$.

(3.8) $\qquad\qquad \mathcal{B} \hspace{1pt}\vert\!\sim^{grounded}_{cr/sc} \phi$ implies $\mathcal{B} \hspace{1pt}\vert\!\sim^{PS}_{cr/sc} \phi$.[8]

3.2.2 From argumentation to dialogical characterisations of non-monotonic reasoning

Given an $AF = (Args, Def)$, argument game proof theories can be defined (e.g., Modgil & Caminada 2009), in which proponent (*PRO*) and opponent (*OPP*) move defeating arguments against each other. The idea is that *PRO* starts by moving an argument X and then *OPP* and *PRO* alternate in moving defeating arguments against each other, by reference to the arguments and defeats in the given AF. If *PRO* successfully defends against all defeats by *OPP*, then *PRO* has shown that there is an admissible extension of the AF that contains X, and so is said to win the game. Restrictions on the players' moves then guarantee that a game is finite, and that the admissible extension is a subset of the grounded extension (in the grounded game) or a preferred extension (in the preferred game). Thus, a game won by *PRO* establishes the conclusion ϕ of X is an argumentation-based credulous, respectively sceptical, *nm* inference from the underlying belief base $\mathcal{B}$, under the preferred, respectively grounded, semantics. Given the aforementioned correspondence results equating the argumentation based inferences from a belief base with the inferences obtained directly from the belief base in various non-monotonic logics, one thus has recourse to argument game proof theories for use by an individual agent (i.e., *OPP* is an imaginary interlocutor) establishing inferences from her belief base in these non-monotonic logics.

---

[8]This result is shown in a forthcoming paper: (van Berkel *et al.*, 2022).

Argument game proof theories can then be generalised and adapted to yield dialogical formalisations of distributed non-monotonic reasoning (Prakken, 2005; Fan & Toni, 2014; Modgil, 2017b), thus paving the way for applications scaffolding inter-agent joint deliberation (as alluded to at the beginning of Section 3.2). The idea is that multiple agents can exchange locutions — not just arguments but claims, questions etc. — that conform to protocols generalising the aforementioned restrictions in games. The protocols and evaluation of the resultant *graph of locutions* that relate locutions based on whether they challenge *or support* each other, are defined so that:

> the status of a communicated claim α is evaluated as "winning" in the dialogue graph iff α is the conclusion of an argument in an extension of the AF defined on the basis of the *belief base consisting of all that has thus far been asserted in the dialogue*[9]

and so, given the above-mentioned correspondences between argumentation based inferences and non-monotonic consequence relations:

> a dialogue concludes in favour of α iff α is non-monotonically inferred from the *belief base consisting of all that has thus far been asserted in the dialogue*

Notice, that under certain assumptions (in particular that the agents are honest and exhaustive) one can aim at results in which "belief base consisting of all that has thus far been asserted in the dialogue" is substituted by "belief base consisting of the union of the belief bases of the dialogue's participating agents", so yielding a more comprehensive account of distributed non-monotonic reasoning.[10]

The research programme aiming at these dialogical characterisations of non-monotonic reasoning reveals a somewhat radical departure from the conventional GOFAI focus on logics for single agent reasoning. In addition to the paradigmatic case of an individual agent reasoning non-monotonically from their *private* static belief base, the focus is on formalising non-monotonic reasoning in terms of the dialogical exchange of locutions. Ratio-

---

[9]In contrast to the case where argument games are defined over AFs defined by a static belief base as in the paradigm case of single agent reasoning.

[10]dialogue protocols that enable one to establish these more comprehensive correspondence results reamin to formalised.

nal logic-based prescriptions for real-world defeasible reasoning — reasoning that needs to accommodate uncertainty and arbitration amongst conflicting beliefs and decision options and is thus better served by pooling the epistemic resources of multiple information sources, *qua* agents — are cashed out in terms of constraints on the contents of speech acts (locutions), and when one speech act is a legitimate reply to another. Indeed, as suggested earlier, the need for such distributed communicative accounts of non-monotonic reasoning is all the more pressing given the increasingly machine learning dominated AI landscape. Moreover, generalisation to a distributed setting sees a departure from the traditional focus on an individual agent's proof theory and its soundness and completeness w.r.t. a model theoretic structure; rather, the focus is on dialogical proof theories,[11] that are sound and complete w.r.t. the available *publically* asserted information (together with the contents of interlocutors' belief bases in the case of the generalised results alluded to above).

In order to practically realise such accounts of distributed non-monotonic reasoning that are suitable for real-world agents, one first needs to develop argumentative formalisations of non-monotonic reasoning that are *rational under resource bounds*.

3.3 Dialectical classical logic argumentation: Rationality under resource bounds

We now propose normal depth-bounded proofs as suitable candidates for classical logic arguments, that, together with a novel development of classical logic argumentation that accommodates real-world features of dialectical reasoning, yield argumentative (and hence dialogical) characterisations of rational resource-bounded non-monotonic reasoning.

3.3.1 *K*-depth classical logic arguments

Standard accounts of classical logic argumentation (such as Amgoud & Cayrol 2002; Gorogiannis & Hunter 2011; Modgil & Prakken 2013) —

---

[11]Note that the dialogical rehabilitation of logic — harking back to the Greek and Scholastic traditions — has also been advocated for *deductive* and, in particular, classical logic (Novaes, 2020).

which henceforth are collectively referred to as "*Cl-Arg*" — typically formalise arguments as tuples (Δ, ϕ) and so leave implicit the specific proof theoretic means by which a conclusion is inferred from premises. Indeed, tuples (Δ, ϕ) might better be conceived of as "argument schemata" that are concretely realised as "arguments proper" through provision of specific proof theories. Moreover, the persuasive force of a non trivial argument is, arguably, partly dependent on whether valid reasoning steps have been employed in inferring the argument's conclusion. For example, a mathematical argument claiming that a certain theorem follows from the Euclidean axioms would be incomplete and entirely unpersuasive without explicit representation of the proof steps involved. It is the means by which the conclusion is obtained that renders the argument understandable, and furthermore, one would not want to rely on the recipient expending resources to reconstruct the proof from assumptions to conclusion. Indeed, the requirement that arguments are transparent and readily understandable is postulated as a key guideline for any practical application (Dung *et al.*, 2010).

We thus propose natural deduction proofs as suitable candidates for proof theoretic explications of propositional classical logic arguments. Natural deduction explications of classical reasoning — as the name suggests — are considered more perspicuous as compared with semantic tableaux or Gentzen type sequent calculi. Chapter 2's *C-intelim* proofs are a particularly suitable candidate, since as discussed in (D'Agostino *et al.*, 2020) standard natural deduction may be natural from the point of view of intuitionistic logic, but it is not so natural from the point of view of classical logic. However *C-intelim* is really natural for classical logic because the introduction and elimination rules closely reflect the classical meaning of the logical operators (i.e., their truth-table interpretation) and the way in which these are used in classical proofs.

Furthermore, we can exploit the notion of agents who, by virtue of having bounded resources and thus falling short of the ideal of logical omniscience, can still be said to be rational when constructing *C-intelim* arguments to a given k-depth. Recall that each increase in depth intuitively equates with an increase in the inferential capabilities of real-world agents; that is to say, the cognitive capabilities of humans to reason hypothetically by introducing virtual information that augments the information to hand (as illustrated with the sudoku examples in the prequel), as well as computa-

3.3. DIALECTICAL CLASSICAL LOGIC ARGUMENTATION

tional agents whose capabilities increase stepwise, in line with each increase in the depth to which the use of virtual information reveals information implicitly contained in the actual information possessed. It is this intuitive equating of inferential capabilities with the nested introduction of virtual information that suggests that agents reasoning to a given depth can be said to be *satisficers* seeking a satisfactory, rather than an optimal, solution: they satisfy a *minimal* notion of rationality[12] whereby within any given bound (i.e., up to depth k) *all* inferences from the actual information possessed, together with the virtual information introduced, are accessible. The informational semantics for *C-intelim* further substantiates this practical notion of minimal rationality: an agent is rational to the extent that her beliefs are "informationally closed" up to a certain fixed depth: she possesses all the information that is practically available to her (with a bounded use of virtual information, reflecting her computational limitations) and with which she can operate (recall the discussion in Chapter 1). This insight into what it means to be minimally rational can be appreciated if one conceives of an alternative formulation of bounded reasoning in which, within a given putative bound, the inferences accessible to an agent are *not* informationally closed, and so according to our proposal would not be minimally rational.

3.3.2 Non-monotonic reasoning, argumentation and rationality

The above observations can be seen as relating exclusively to monotonic reasoning; at issue is the extent to which an agent can be said to be rational, given the resources expended on deductively reasoning from a given set of consistent premises that are taken at face value. The premises are not subject to challenge, and the focus is exclusively on the connection between premises and conclusions, not on the nature or plausibility of the premises or conclusions (Novaes 2020 refers to this as the *bracketing belief* requirement of deductive reasoning). Of course, real-world agents reasoning non-monotonically are subject to further requirements that arise when needing to account globally for information that may challenge and thus instigate withdrawal of the premises and/or conclusion of a given inference/argument. The question then arises as to whether the outcomes of agent reasoning can be said to respect criteria of rationality, despite having insufficient resources

---

[12]In a sense distinct from that stipulated by (Cherniak, 1986b).

to exhaustively check for all possible inferences or arguments that may instigate withdrawal.[13] In particular, a number of so called "rationality postulates" have been extensively studied in the argumentation literature. Of particular note are the consistency (Caminada & Amgoud, 2007) and non-contamination Caminada *et al.* (2012) postulates. The former states that no complete extension (i.e., maximal set of *conflict free acceptable* — i.e., *admissible* — arguments) contains arguments with contradictory conclusions:

Consistency: For any complete extension E of an AF $(Args, Def)$: $\neg \exists X, Y \in E$ such that $\text{conc}(X) = -\text{conc}(Y)$.

Recalling the argumentation-based consequence relations defined by a belief base $\mathcal{B}$ (Equations 3.5 and 3.6), non-contamination expresses a minimal rational criterion for relevance: no argumentation-based non-monotonic inferences obtained from a belief base $\mathcal{B}$ should be withdrawn if $\mathcal{B}$ is expanded by syntactically disjoint[14] beliefs $\mathcal{B}'$ (the relation of being syntactically disjoint is denoted by $||$):

Non-contamination: If $\mathcal{B} \mathrel{\vert\!\sim}_{sc/cr}^{complete} \phi$ then $\forall \mathcal{B}'$ such that $\mathcal{B} || \mathcal{B}'$: $\mathcal{B} \cup \mathcal{B}' \mathrel{\vert\!\sim}_{sc/cr}^{complete} \phi$.

Given a belief base $\mathcal{B}$, standard accounts of classical logic argumentation (*Cl-Arg*) assume logical omniscience — that the set *Args* in $(Args, Def)$ consist of *all* arguments defined by $\mathcal{B}$ — which is clearly an impractical assumption for real-world resource-bounded agents, and moreover, is tacitly assumed when showing that the consistency rationality criterion is satisfied. The issue is best appreciated if we switch our attention to admissible extensions. Recall that in the argument games and dialogues reviewed in Section 3.2.2, it suffices that *PRO* establishes membership of an argument (respectively the status of a claim) in an admissible extension. One would therefore want a criterion of consistency to apply to admissible extensions,

---

[13] In the maxiconsistent paradigm only the premises can be challenged, given that a conclusion cannot be withdrawn while maintaining that the premises hold.

[14] See (D'Agostino & Modgil, 2018a) for the precise definition of when two sets of first-order formulae are syntactically disjoint; when they share no predicate or function (including constant) symbols (in the propositional case this equates to no shared propositional atomic formulae).

3.3. DIALECTICAL CLASSICAL LOGIC ARGUMENTATION

and hence we focus on the consistency of premises in an admissible extension, in the sense that the premises moved by *PRO* should not entail $\bot$.

Premise Consistency: For any admissible extension E of an AF $(Args, Def)$: $\bigcup_{X \in E} \text{prem}(X) \nvdash \bot$.

Notice that under the assumption of logical omniscience, satisfaction of *Premise Consistency* is equivalent to satisfaction of *Consistency*.[15]

Consider now the set of arguments E in Figure 3.2i that *PRO* moves in the course of defending an argument. Clearly E violates *Premise Consistency* and so one would want that *OPP* moves arguments to show that E is not admissible. The ability to do so assumes that resources suffice to construct all arguments defined by the premises in A, B and C; in particular X, Y and Z, which respectively attack A, B and C. To see why, *suppose at least one of these attacks succeed as a defeat*. Then any argument in E that defends against this defeat, by defeating X or Y or Z, must target one of p, $p \to q$ or $p \to \neg q$, which in turn means that the defending argument in E must then also target and defeat A or B or C, and so E would not be conflict free and hence not admissible. Thus, for *OPP* to challenge the admissibility of E on the grounds that *PRO* commits to inconsistent premises, *OPP* must construct and deploy arguments that force *PRO* to contradict himself by defeating one of his own arguments.

Note that the above italicised supposition — *that one of the attacks succeeds as a defeat* — is not entirely innocent, since one then requires that the preference relation be such that either $X \not\prec (\{p\}, p)$ or $Y \not\prec (\{p \to q\}, p \to q)$ or $Z \not\prec (\{p \to \neg q\}, p \to \neg q)$ (in which case $\prec$ is said to be *reasonable* — see (Modgil & Prakken, 2013) for details). Moreover, there is a tacit assumption of logical omniscience. Suppose *OPP* is capable of constructing only 0-depth arguments, in which case she can construct Y and Z, but has insufficient resources to construct X, which is a 1-depth argument.[16] Then, if $Y \prec (\{p \to q\}, p \to q)$ and $Z \prec (\{p \to \neg q\}, p \to \neg q)$, and since

---

[15] If $X, Y \in E$ a complete extension, and $\text{conc}(X) = -\text{conc}(Y)$, then clearly $\text{prem}(X) \cup \text{prem}(Y) \vdash \bot$ and if $\bigcup_{Z \in E} \text{prem}(Z) \vdash \bot$ it can straightforwardly be shown that $\exists X, Y \in E$, $\text{conc}(X) = -\text{conc}(Y)$.

[16] Referring to the rules of C-intelim tableaux in Section 2.1.4, assume (by RB) q and $\neg q$, and from $q, p \to \neg q$ (by $\to E_2$) $\neg p$, from $\neg q, p \to q$ (by $\to E_2$) $\neg p$, concluding $\neg p$ on discharge of q and $\neg q$.

OPP cannot then defeat A on $(\{p\}, p)$ with X, *PRO* is able to defend her arguments with an admissible extension that violates *Premise Consistency*. Indeed, for any AF consisting only of k-depth arguments, one can always devise examples to show that although the premises in a set E of k-depth arguments are k-depth inconsistent, E is admissible. The preceding example exemplifies this for $k = 0$. (Note that a similar analysis applies to showing *Consistency* for complete extensions).

Satisfaction of *Non-contamination* also makes unacceptable demands on the resources available to agents. For each constructed argument, one must not only check consistency (*Prem-con*) of the premises but also that they are subset-minimal (*Prem-rel*) — i.e., conditions 3) and 4) respectively in Equation 3.1 — where the latter is a problem in the second level of the polynomial hierarchy (Eiter & Gottlob, 1995) (in the worst case, for every constructed argument (Δ, ϕ), one needs to verify that $\forall \Delta' \subset \Delta, \Delta' \nvdash \alpha$). However, *Prem-con* and *Prem-rel* ensure satisfaction of *Non-contamination*.

Firstly, suppose *Prem-con* is not enforced. Let $\mathcal{B} = \{a, c \to \neg a\}$ and so

$$\mathcal{B} \mathrel{\vert\!\sim}_{sc}^{grounded} a.$$

Let $\mathcal{B}' = \{b, \neg b\}$, and so $A = (\{a\}, a)$, $C = (\{b, \neg b\}, c)$ and $B = (\{b, \neg b, c \to \neg a\}, \neg a)$ are arguments defined by $\mathcal{B} \cup \mathcal{B}'$. Then, assuming $B \not\prec A$, B defeats A and it is easy to see that the minimal complete — i.e., grounded — extension does not include A.[17] Hence:

$$\mathcal{B} \cup \mathcal{B}' \mathrel{\vert\!\not\sim}_{sc}^{grounded} a.$$

Secondly, suppose *Prem-rel* is not enforced. Let $\mathcal{B} = \{a, \neg a\}$, $A = (\{a\}, a)$ and $B = (\{\neg a\}, \neg a)$, and $B \prec A$. Hence A defeats B, B does not defeat A, and the grounded extension includes A:

$$\mathcal{B} \mathrel{\vert\!\sim}_{sc}^{grounded} a.$$

Now, let $\mathcal{B}' = \{c\}$ and $C = (\{c, \neg a\}, \neg a)$ and suppose $C \not\prec A$ and $A \not\prec C$. Now as well as A defeating B, C and A defeat each other, the grounded

---

[17]Note also the assumption that $(\{b\}, b) \not\prec (\{\neg b\}, \neg b)$ and $(\{\neg b\}, \neg b) \not\prec (\{b\}, b)$ so that neither of these arguments are in the grounded extension and neither can defeat and so defend against B's defeat on A.

3.3. DIALECTICAL CLASSICAL LOGIC ARGUMENTATION

extension is $\emptyset$ and

$$\mathcal{B} \cup \mathcal{B}' \not\vdash_{sc}^{grounded} a.$$

3.3.3 Dialectical acceptability and k-depth argumentation frameworks

Recall now (Chapter 2) the notion of a *normal* C-intelim proof, which can be generated by straightforward restrictions on the applications of the rules that (i) limit the choice of the RB-formula of an RB-application to a tractable space defined by the assumptions and the conclusion and (ii) avoid obvious redundancies in the construction of a proof. Specifically, normal C-intelim proofs for unsigned formulae enjoy the (weak) subformula property, which makes their construction amenable to algorithmic treatment, and moreover, exclude proofs that incorporate redundant syntactically disjoint and consistent premises (see Proposition 2.3.10); in particular arguments such as C above — referred to (see Definition 2.3.4) as "redundantly contaminated" or "R-contaminated" arguments — which incorporates the redundant syntactically disjoint c in the premises. On the other hand, *normal* C-intelim proofs can yield arguments of the form $B = (\{b, \neg b\}, \neg a)$.

Hence, dropping the computationally impractical *Prem-rel* check on arguments' premises, and confining arguments to *normal* C-intelim proofs, one can avoid violation of *Non-contamination* arising from the use of R-contaminated arguments. However, dropping the *Prem-con* check on arguments' premises may still result in violating *Non-contamination*. On the other hand, the idea that an agent introspects on the consistency of her argument's premises is not borne out by real-world dialectical practice. Rather, the inconsistency of an argument's premises is demonstrated dialectically, as a special case of a typical dialectical move whereby interlocutors distinguish their own premises, namely those that they accept as true, from the premises that their opponent commits to and that they want to criticise:

> "on the basis of the premises I regard to be true, and supposing for the sake of argument what you regard to be true, then I can show some conclusion that contradicts one of your premise".

Thus, if argumentation is to be generalisable to distributed accounts of non-monotonic reasoning, one would ideally accommodate this epistemo-

CHAPTER 3. RATIONAL NON-MONOTONIC REASONING

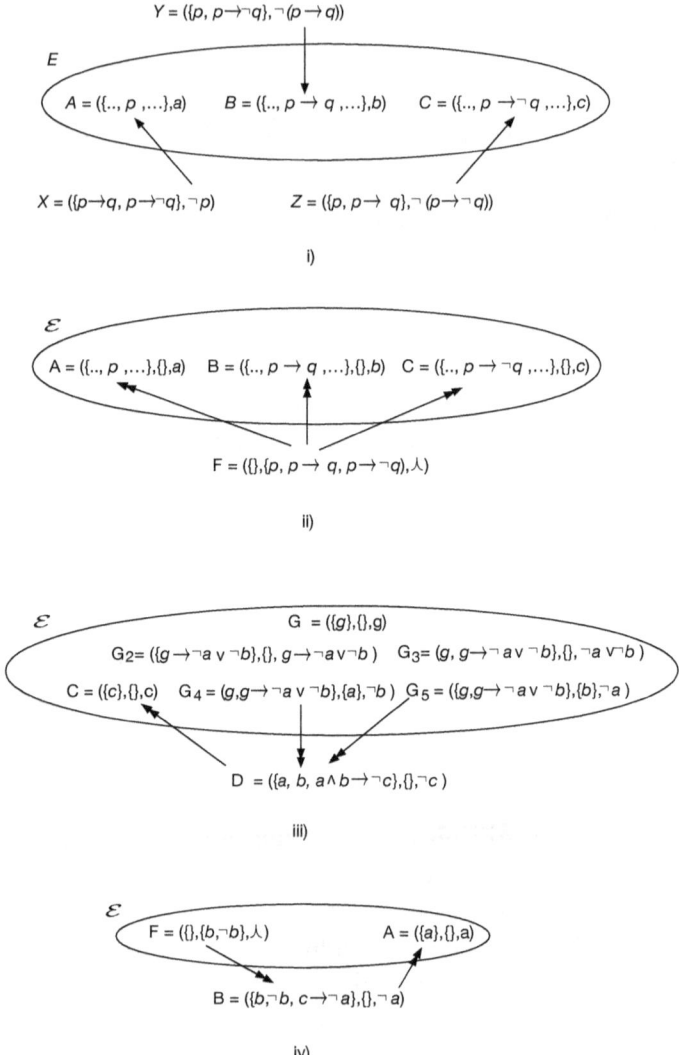

Figure 3.2: In i) single headed arrows denote attacks amongst arguments defined in the standard classical logic argumentation paradigm *Cl-Arg*. In ii), iii) and iv) double headed arrows denote defeats in *Dialectical* Classical Logic Argumentation.

3.3. DIALECTICAL CLASSICAL LOGIC ARGUMENTATION

logical distinction — between committed and supposed premises — when evaluating and moving arguments. Now, in the case that an argument's premises are inconsistent, this can then be challenged dialectically:

> "Supposing only your premises, I can show that you contradict yourself".

For example, the improper 0-depth argument $B = (\{b, \neg b\}, \neg a)$ would intuitively be challenged by an argument that commits to no premises but supposes only the premises of B to conclude (at depth 0) that these premises are inconsistent (i.e., entail $\curlywedge$). Recall, that improper proofs can be turned in linear time into a non contaminated proof of $\curlywedge$.

Indeed, this latter dialectical use of an argument is what one might refer to as the "Socratic move", in which an interlocutor highlights her opponent's commitments to inconsistent premises in the course of a dialogue.[18] In the example in Figure 3.2i, the premises $p, p \to q, p \to \neg q$ in E are 0-depth inconsistent, and *OPP* should intuitively challenge the admissibility of E by moving an argument entailing $\curlywedge$ from these premises; premises that are only supposed for the sake of argument given that they are committed by *PRO*. *OPP*'s argument would then target any *PRO* argument in E that makes use of one these culpable premises so as to dialectically demonstrate that *PRO* contradicts herself.

In summary, we propose accommodating resource-bounded agents in virtue of the study of k-depth Argumentation Frameworks (as reported on in D'Agostino & Modgil 2018a) in which the arguments are k-depth normal C-intelim proofs (whose premises are not validated for consistency or subset minimality), and such that evaluating the dialectical acceptability of arguments explicitly acknowledges the epistemological distinction characteristic of real world dialectical practice.

Definition 3.3.1. *Let $\mathcal{B}$ be any finite belief base of propositional classical wff. Then for some $k \in \mathbb{N}$, $(\mathcal{A}_k, \mathcal{C}_k)$ is a k-depth argumentation framework (AF_k) defined by $\mathcal{B}$, where $\mathcal{A}_k$ is the set (Δ, ϕ) of normal C-intelim$_k$*

---

[18]For example, in his *Dialogues Concerning Two New Sciences* (Vlastos, 1982), Galileo presents a famous refutation of Aristotle's theory of falling bodies, in the form of a dialogue between their respective alter-egos Salviati and Simplicio. Salviati demonstrates that the premises of Simplicio's arguments justifying that heavier bodies fall faster than lighter bodies, lead to a contradiction (i.e., are inconsistent).

proofs of ϕ depending on Δ and the binary attack relation $\mathcal{C}_k \subseteq 2^{\mathcal{A}_k} = \{(X,Y)|X,Y \in \mathcal{A}_k, \text{conc}(X) = -\alpha, \alpha \in \text{prem}(Y)\}$ (recall Section 3.2.1).

The distinction ubiquitous in dialectical practice, between the premises one commits to and those "supposed for the sake of argument" that are committed by a (possibly imaginary) interlocutor is manifested when constructed arguments in $\mathcal{A}_k$ are deployed dialectically; that is, when defining the notion of dialectical defeats that in turn determines membership of arguments in extensions under the Dung semantics. Thus an epistemic variant of a (k-depth) normal C-intelim proof of ϕ depending on Δ, is one that partitions Δ into committed and supposed premises.

Definition 3.3.2. *Let $X = (\Delta, \phi)$ be an argument as defined in Definition 3.3.1. Then $\mathsf{X} = (\Sigma, \Gamma, \phi)$ is an epistemic variant of X iff $\Delta = \Sigma \cup \Gamma$ and $\Sigma \cap \Gamma = \emptyset$.*

Notation 3.3.3. *Let $\mathsf{X} = (\Sigma, \Gamma, \phi)$ be an epistemic variant of X:*

- $\text{conc}(\mathsf{X}) = \text{conc}(X) = \phi$;

- $\text{Com}(\mathsf{X})$ *denotes the committed premises (commitments) Σ of X, and $\text{Sup}(\mathsf{X})$ denotes the supposed premises (suppositions) Γ of X;*

- $|X|$ *denotes the set of all epistemic variants of X, $|E|$ denotes $\bigcup_{X \in E} |X|$ where $E \subseteq \mathcal{A}_k$ and $\text{Com}(|E|)$, $\text{Sup}(|E|)$ and $\text{conc}(|E|)$ respectively denote $\bigcup_{\mathsf{X} \in |E|} \text{Com}(\mathsf{X})$, $\bigcup_{\mathsf{X} \in |E|} \text{Sup}(\mathsf{X})$ and $\bigcup_{\mathsf{X} \in |E|} \text{conc}(\mathsf{X})$.*

- $\mathcal{S}, \mathcal{E}$ *denote subsets of epistemic variants. That is, $\mathcal{S} \subseteq |\mathcal{A}_k|$, $\mathcal{E} \subseteq |\mathcal{A}_k|$.*

In general, when establishing whether $\mathsf{X} = (\Sigma, \Pi, \alpha)$ is defended by (acceptable w.r.t.) a set $\mathcal{E}$, it is only the committed premises Σ that can be targeted. An attack by $\mathsf{Y} = (\Delta, \Gamma, \alpha)$ on $\beta \in \Sigma$ is contingent on the suppositions Γ of Y being commitments in $\mathcal{E}$ and X. Intuitively:

> "given that I commit to Δ and supposing for the sake of argument your commitments Γ in $\mathcal{E} \cup \{\mathsf{X}\}$, I can construct an argument Y that challenges your premise $\beta \in \Sigma$sake".

3.3. DIALECTICAL CLASSICAL LOGIC ARGUMENTATION 129

Such an attack succeeds as a defeat only if the elementary argument corresponding to β (i.e.,$(\{\beta\}, \beta)$) is not strictly preferred to Y. An argument of the form $Y = (\Delta, \Gamma, \lambda)$ can be used to challenge X by arguing that the premises Γ committed in $\mathcal{E} \cup \{X\}$ (i.e., $\Gamma \subseteq \text{Com}(\mathcal{E} \cup \{X\})$), together with Δ, are inconsistent. X should only then be targeted if at least one of its committed premises $\beta \in \text{Com}(X)$ is in Γ and so is culpable in contributing to the inconsistency. Again, Y can be moved as a defeat subject to $(\{\beta\}, \beta)$ not being strictly preferred to Y. However, if $\Delta = \emptyset$ then Y dialectically demonstrates that the proponent of X has committed to premises Γ in $\mathcal{E} \cup \{X\}$ that are inconsistent. To prefer that one commits to inconsistent premises is clearly incoherent, and so this defeat is *not* preference dependent. Finally, $Z \in \mathcal{E}$ can defend X by defeating Y, while supposing any of Y's commitments (i.e., $\text{Sup}(Z) \subseteq \text{Com}(Y)$).

Definition 3.3.4. *Let $(\mathcal{A}_k, \mathcal{C}_k)$ be a k-depth AF defined by $\mathcal{B}$, and $\prec$ a strict partial ordering over $\mathcal{A}_k$. Let $\mathcal{E} \subseteq |\mathcal{A}_k|$, $Y = (\Delta, \Gamma, \phi) \in |Y|$, $X = (\Pi, \Sigma, \psi) \in |X|$, $X, Y \in \mathcal{A}_k$.*

1. *if $\phi \neq \lambda$, then Y defeats X w.r.t. $\mathcal{E}$, denoted $Y \Rightarrow_\mathcal{E} X$, iff*

    (a) $(Y, X) \in \mathcal{C}_k$ on $X' = (\{\beta\}, \beta)$, $\beta \in \text{Com}(X)$ and $Y \not\prec X'$;

    (b) $\Gamma \subseteq \text{Com}(\mathcal{E} \cup \{X\})$;

    We say that Y defeats X on $X' = (\{\beta\}, \beta)$ or Y defeats X on β.

2. *if $\phi = \lambda$, then Y defeats X w.r.t. $\mathcal{E}$, denoted $Y \Rightarrow_\mathcal{E} X$, iff*

    (a) $\Gamma \cap \text{Com}(X) \neq \emptyset$ and $\Gamma \subseteq \text{Com}(\mathcal{E} \cup \{X\})$;

    (b) *either* $\Delta = \emptyset$ *or* $\forall \beta \in \Gamma \cap \text{Com}(X), Y \not\prec (\{\beta\}, \beta)$

    We say that Y defeats X on $X' = (\{\beta\}, \beta)$ or Y defeats X on β, where $\beta \in \Gamma \cap \Pi$.

Definition 3.3.5. *Let $\mathcal{E} \subseteq |\mathcal{A}_k|$:*

- $X \in |X|$ *is acceptable w.r.t. $\mathcal{E}$ iff $\forall Y$ s.t. $Y \Rightarrow_\mathcal{E} X$, $\exists Z \in \mathcal{E}$ s.t. $Z \Rightarrow_{\{Y\}} Y$.*

- $\mathcal{E} \subseteq |\mathcal{A}_k|$ *is conflict free iff $\neg \exists X, Y \in \mathcal{E}$ s.t. $Y \Rightarrow_\mathcal{E} X$.*

- Let $\mathcal{E} \subseteq |\mathcal{A}_k|$ be conflict free. Then $\mathcal{E}$ is: admissible iff $\forall \mathsf{X} \in \mathcal{E}$, X is acceptable w.r.t. $\mathcal{E}$; a complete extension iff $\mathcal{E}$ is admissible and $\forall \mathsf{X} \in |\mathcal{A}_k|$, X is acceptable w.r.t. $\mathcal{E}$ implies $\mathsf{X} \in \mathcal{E}$; the grounded extension iff $\mathcal{E}$ is the minimal under set inclusion complete extension; a preferred extension iff $\mathcal{E}$ is a maximal under set inclusion complete extension; a stable extension iff $\forall \mathsf{Y} \in |\mathcal{A}_k| \setminus \mathcal{E}$, $\exists \mathsf{X} \in \mathcal{E}$ s.t. $\mathsf{X} \Rightarrow_{\{\mathsf{Y}\}} \mathsf{Y}$.

Remark 3.3.6. *Notice that if $\Delta = \emptyset$ in the epistemic variant $\mathsf{X} = (\Delta, \Gamma, \phi)$ then X cannot be defeated given its empty commitments, and so X is clearly acceptable w.r.t. any set of arguments, and hence is said to be* unassailable.

To recap, we assume a Dung AF $(\mathcal{A}_k, \mathcal{C}_k)$ consisting of arguments that are normal C-intelim proofs related by binary attacks. The premises of these arguments need *not* be checked for subset minimality or consistency. It is only in the dialectical use of these arguments, when determining the acceptability of arguments, that the epistemic commitment/supposition distinction is deployed through the use of epistemic variants. Once the *dialectical* extensions — subsets of $|\mathcal{A}_k|$ — under the various semantics are defined as in Definition 3.3.5, only the conclusions of *unconditional* arguments that commit to *all* their premises, identify the claims supported by the extensions. We then identify the extensions of $(\mathcal{A}_k, \mathcal{C}_k)$ — subsets of $\mathcal{A}_k$ — by reference to the unconditional arguments in the dialectical extensions.

Definition 3.3.7. *Let $(\mathcal{A}_k, \mathcal{C}_k)$ be a k-depth argumentation framework defined by $\mathcal{B}$, and $\prec$ a strict partial ordering over $\mathcal{A}_k$. For $s \in \{$ admissible, complete, grounded, preferred, stable $\}$, let $\mathcal{E} \subseteq |\mathcal{A}_k|$ be a dialectical s-extension of $(\mathcal{A}_k, \mathcal{C}_k)$ as defined in Definition 3.3.5. Then $E = \{(\Delta, \phi) | (\Delta, \emptyset, \phi) \in \mathcal{E}\}$ is an s-extension of $(\mathcal{A}_k, \mathcal{C}_k)$.*

Let $E_1, \ldots, E_n$ be the s-extensions of $(\mathcal{A}_k, \mathcal{C}_k)$. Then the argumentation defined sceptical, respectively credulous, consequence relations are defined as follows:

$$\mathcal{B} \mathrel{\vert\!\sim}^s_{sc} \phi \text{ iff } \phi \in \bigcap_{i=1}^{n} \mathsf{conc}(E_i)$$

$$\mathcal{B} \mathrel{\vert\!\sim}^s_{cr} \phi \text{ iff } \exists E_i, \phi \in \mathsf{conc}(E_i)$$

Recall now the interpretation of propositional letters and arguments in Example 3.2.1 and that the non-monotonic (sceptical) consequence relation

3.3. DIALECTICAL CLASSICAL LOGIC ARGUMENTATION

defined by Preferred Subtheories (*PS*) (Brewka, 1989) yields:

$$\mathcal{B} \mathrel{|\!\sim}^{PS}_{sc} c$$

whereas under the *Cl-Arg* characterisation of *PS*:

$$\mathcal{B} \mathrel{|\!\not\sim}^{grounded}_{sc} c$$

since $C = (\{c\}, c)$ is *not* in the grounded extension.

To see why, observe that in Figure 3.1c), $D_2 = (\{a, b, (a \wedge b) \to \neg c\}, \neg c)$ defeats C, but in order to defend against this defeat, either $G_4 = (\{g, a, g \to \neg a \vee \neg b\}, \neg b)$ or $G_5 = (\{g, b, g \to \neg a \vee \neg b\}, \neg a)$, and hence $A = (\{a\}, a)$ or $B = (\{b\}, b)$, must be in the grounded extension E_G. But neither A or B are in E_G, since the budgetary constraints preclude attendance at conference a *and* b — $G_3 = (\{g, g \to \neg a \vee \neg b\}, \neg a \vee \neg b) \in E_G$ — and given no information to arbitrate in favour of attending a or b, one sceptically commits to attendance at neither.

Now, if G_3 were able to attack and defeat D_2 on a *subset* $\{a, b\}$ of prem(D_2), then C would indeed be defended and so a member of E_G. However, if the standard approach to classical logic argumentation were to allow what (Gorogiannis & Hunter, 2011) call "undercut attacks":

> X attacks Y if X's conclusion is classically equivalent to the negation of any subset of Y's premises

then not only does this allow for counter-examples to the consistency postulate (see Gorogiannis & Hunter 2011), but desiderata for practical realisations of argumentation are undermined; in particular that whether one argument attacks another should be "directly inspectable" (i.e., identifiable in linear time) (Dung *et al.*, 2010).

However, in Dialectical *ClArg* (D'Agostino & Modgil, 2018a,b) one can effectively simulate undercuts through use of suppositions, while preserving consistency and the requirement that attacks be directly inspectable. Suppose:

$$X = (\Delta, \emptyset, \phi) \in |\mathcal{A}_k| \text{ and } Y = (\Pi, \Sigma, \gamma) \in |\mathcal{A}_k| \text{ where } \phi \dashv\vdash_k \neg \bigwedge \Theta \text{ for some } (\Theta = \{\alpha_1, \ldots, \alpha_m\}) \subseteq \Pi.$$

Then for $i = 1 \ldots m$: $\exists X_i = (\Delta, \Theta \setminus \{\alpha_i\}, \neg \alpha_i) \in |\mathcal{A}_k|$ that targets $\alpha_i \in \Theta$, *where none of the premises $\Theta \setminus \{\alpha_i\}$ need be committed.*

To illustrate, consider the set $\mathcal{E}$ of 0-depth dialectical arguments in Figure 3.2iii. $\mathcal{E} \subseteq |\mathcal{A}_0|$ is a subset of the *dialectical* grounded extension as defined in Definition 3.3.5, since the 0-depth arguments G_4 and G_5 that respectively cite a and b *only as suppositions*, defeat D and so defend C. Hence, defining the grounded extension in terms of the unconditional arguments in $\mathcal{E}$ (Definition 3.3.7):

$$\{C = (\{c\}, c), G_1 = (\{g\}, g), G_2 = (\{g \to \neg a \vee \neg b\}, g \to \neg a \vee \neg b), G_3 = (\{g, g \to \neg a \vee \neg b\}, \neg a \vee \neg b)\} \subseteq \mathcal{A}_k$$

is a subset of the grounded extension, and so we recover $\mathcal{B} \hspace{0.2em}\vert\hspace{-0.5em}\sim_{sc}^{grounded} c$. That is to say, *Dialectical ClArg* appropriately yields sceptical inferences that are excluded by *ClArg*.

Premise Consistency and *Consistency* are satisfied by k-depth AFs. Recall the *Cl-Arg* example illustrated in Figure 3.2i. *Premise Consistency* is violated if $Y \prec B$, $Z \prec C$ and resources do not suffice to construct X (Y and Z are 0-depth arguments and X is a 1-depth argument). However, consider the set of unconditional epistemic variants $\mathcal{E}$ in Figure 3.2ii, given:

$$A = (\{.., p, ..\}, a), B = (\{.., p \to q, ..\}, b), C = (\{.., p \to \neg q, ..\}, c) \subseteq \mathcal{A}_0 \text{ where } (\mathcal{A}_0, \mathcal{C}_0) \text{ is a 0-depth } AF \text{ defined by a belief base } \mathcal{B},$$

$\mathcal{E} \subseteq |\mathcal{A}_0|$ cannot be a subset of a dialectical admissible extension (hence $E \subseteq \mathcal{A}_0$ in Figure 3.2i cannot be admissible) since the 0-depth

$$\mathsf{F} = (\{\}, \{p, p \to q, p \to \neg q\}, \curlywedge)$$

defeats each of A, B and C on (respectively) the premises p, $p \to q$ and $p \to \neg q$ that are committed in $\mathcal{E}$ and that entail $\curlywedge$ at depth 0. F is *unassailable* (i.e., does not commit to any premises) and hence cannot be defeated, and so the defeats on A, B and C cannot be defended by any argument included in $\mathcal{E}$. In general, the following results hold (see D'Agostino & Modgil 2018a for formal proofs):

3.3. DIALECTICAL CLASSICAL LOGIC ARGUMENTATION

Premise Consistency: For any admissible extension E of a k-depth AF: $\bigcup_{X \in E} \text{prem}(X) \nvdash_k \lambda$.

Consistency: For any admissible extension E of a k-depth AF: $\neg \exists X, Y \in E$ such that $\text{conc}(X) = -\text{conc}(Y)$ and $\neg \exists Z \in E$ such that $\text{conc}(Z) = \lambda$.

Non-contamination is also satisfied by k-depth AFs, despite neither of the *Prem-rel* and *Prem-con* checks being enforced on arguments' premises (see D'Agostino & Modgil 2018a for a formal proof). Firstly, since arguments are normal C-intelim proofs, and so cannot be R-contaminated, then when expanding a base $\mathcal{B} = \{a, \neg a\}$ with $\mathcal{B}' = \{c\}$ (recall the counter-example to satisfaction of *Non-contamination* by *Cl-Arg* in Section 3.3.2), there is no normal C-intelim proof of $\neg a$ — that is to say no argument $(\{c, \neg a\}, \neg a)$ — that depends on c and $\neg a$.

Secondly, when expanding a base $\mathcal{B} = \{a, c \to \neg a\}$ with $\mathcal{B}' = \{b, \neg b\}$ *Non-contamination* is threatened by virtue of the "explosive" argument $B = (\{b, \neg b, c \to \neg a\}, \neg a)$. However, since any such improper C-intelim proof can be turned in linear time into a non contaminated proof of λ from b and $\neg b$, then any k-depth AF that contains B, also contains $F = (\{b, \neg b\}, \lambda)$. Hence, given the "explosively contaminating" epistemic variant

$$\mathsf{B} = (\{b, \neg b, c \to \neg a\}, \{\}, \neg a)$$

that defeats $A \in \mathcal{E}$ (see Figure 3.2iv), one can defend against this defeat with $\mathsf{F} = (\{\}, \{b, b\}, \lambda) \in \mathcal{E}$. Moreover, F is unassailable and hence acceptable w.r.t. any set of dialectical arguments, and so can be included in any such set $\mathcal{E}$.

Finally, note that the correspondence results established for Dialectical *Cl-Arg* and the Preferred Subtheories non-monotonic logic (Equations 3.7 and 3.8), together with the provably rational outcomes obtained for k-depth AFs, implies that (for the propositional case) we obtain a resource bounded rational and tractable formalisation of a non-monotonic logic. Independent of the logic's dialectical characterisation, one can now provide a formalisation of k-depth Preferred Subtheories reasoning, wherein the preferred maximal consistent subsets of a belief base $\mathcal{B}$ are now obtained assuming only $\vdash_k \subseteq \vdash$.

3.4 Conclusions and further remarks

This chapter's application of depth-bounded classical deduction represents the beginning of a programme of research that seeks to rehabilitate the study and application of non-monotonic logics, in the context of a waning of interest in non-monotonic logics that in large part is due to the computational intractability of non-monotonic reasoning, and the increasing focus on machine learning (ML) in contemporary AI. The quantitative ML paradigm necessarily requires augmentation by symbolic AI so as to both render ML reasoning explainable and transparent for human inspection, and to enable machines and human to jointly reason. It is this latter requirement that motivates development of dialogical characterisations of non-monotonic reasoning; characterisations that accommodate real-world modes of dialectical reasoning that yield rational outcomes while eschewing the assumption of logical omniscience.

This chapter's dialectical formalisation of maxiconsistent approaches to non-monotonic reasoning accommodates resource bounded k-depth agents exchanging normal C-intelim proofs that satisfy a minimal notion of rationality — an agent is rational to the extent that her beliefs are "informationally closed"; she possesses all the information that is practically available, augmented by the virtual information introduced up to depth k, and with which she can operate — and that yields rational outcomes when expending finite resources on verifying the legitimacy of a given non-monotonic inference. Specifically, an agent can commit with confidence (that the mutual k-depth consistency of her inferences is preserved) to inferences from $\mathcal{B}$, having verified that the information to hand, augmented by k-depth virtual information, does not yield reasons for withdrawing these inferences. Moreover, her inferences are preserved when augmenting $\mathcal{B}$ with (syntactically disjoint) information that clearly bears no relevance on whether these inferences should be withdrawn.[19]

We point to next steps in the research programme. Firstly, we will further investigate k-depth AFs consisting of depth-bounded natural deduction

---

[19] This result could be expressed as "Relevant Monotony" — If $\mathcal{B} \mathrel{|\!\sim} \phi$ and $\mathcal{B} \| \mathcal{B}'$ then $\mathcal{B} \cup \mathcal{B}' \mathrel{|\!\sim} \phi$ — in addition to the widely studied "Cautious Monotony" (If $\mathcal{B} \mathrel{|\!\sim} \phi$ and $\mathcal{B} \mathrel{|\!\sim} \theta$ then $\mathcal{B} \cup \{\theta\} \mathrel{|\!\sim} \phi$) and "Rational Monotony" (If $\mathcal{B} \mathrel{|\!\sim} \phi$ and $\mathcal{B} \mathrel{|\!\not\sim} \neg\theta$ then $\mathcal{B} \cup \{\theta\} \mathrel{|\!\sim} \phi$) (Strasser & Antonelli, 2019).

3.4. CONCLUSIONS AND FURTHER REMARKS

proofs (i.e., arguments) for full first-order logic as presented in (D'Agostino et al., 2021). Results in (D'Agostino & Modgil, 2018a) imply that such frameworks also yield consistent outcomes, since it can be shown that so long as

(**A1**) there is a k-depth proof of $\curlywedge$ depending on Δ and Γ, given a k-depth proof of ϕ depending on Δ, and a k-depth proof of $-\phi$ depending on Γ

then consistency is satisfied.

Moreover, given the notion of a *normal* k-depth first-order proof, (D'Agostino & Modgil, 2018a) shows that *Non-contamination* is satisfied so long as the following holds:

(**A2**) Given a non-normal k-depth proof of ϕ depending on $\Delta \cup \Gamma$ (i.e., an argument $X = (\Delta \cup \Gamma, \phi)$), such that $\Gamma \| \Delta \cup \{\phi\}$, then either:

1. there exists a k-depth proof of ϕ depending on Δ (i.e., an argument $Y = (\Delta, \phi)$[20] and X is not "stronger" than Y (in other words adding syntactically disjoint assumptions to Y does not entail that for any Z: $Z \not\prec Y$ and $Z \prec X$, or $Y \prec Z$ and $X \not\prec Z$), or;

2. there exists a k-depth proof of $\curlywedge$ depending on Γ.

Secondly, we will build on work in (D'Agostino & Modgil, 2020) that deploys the dialectical evaluation of epistemic variants of arguments that also incorporate defeasible inference rules. In (D'Agostino & Modgil, 2020), arguments are formalised within the so called "*ASPIC+*" framework for argumentation (Modgil & Prakken, 2013), that has been used to establish correspondence results for a wide range of non-monotonic logics; both those within the maxiconsistent *and* defeasible inference rule paradigm. Preliminary results in (D'Agostino & Modgil, 2020) establish satisfaction of all the rationality postulates, while making only minimal assumptions as to the

---
[20]That is to say, Y has been redundantly contaminated by the syntactically disjoint Γ to yield the R-contaminated X.

resources available for constructing arguments (essentially the quoted assumptions **A1** and **A2** above, with the term "argument" substituting for "k-depth proof"). Finally, we will aim at dialogical accounts of non-monotonic reasoning, generalising argument game proof theories for Dialectical *ASPIC+*, so as to yield frameworks for distributed reasoning that are rational under resource bounds.

Chapter 4

Enduring problems in the philosophy of logic

4.1 Is logic analytic?

By and large, philosophers say that a statement is "analytic" when it can be justified by means of conceptual analysis only, and "synthetic" when its justification requires us to process information that is not "given" in the statement. A typical (but not uncontroversial) example of an analytic statement is "no bachelor is married", once we realize that "bachelor" means "unmarried male". On the other hand, "Bob is married" is synthetic in that it cannot be justified by mere analysis of the statement itself. This distinction goes hand in hand with the distinction between *a priori* statements, that do not depend on experience — e.g., "a straight line is the shortest between two points" or "7 + 5 = 12" — and *a posteriori* ones that can be established only by means of experience — e.g., "this book has 250 pages".

The epistemological status of logic has been at the core of a lively debate in the history of the discipline. This debate not only concerned the exact definition of the analytic-synthetic distinction, but also the constitutive features and aims of the discipline of logic. The idea that logic is analytic and *a priori*, which soon became traditional, finds its most fertile ground in the philosophical movement known as "logical empiricism", which flourished in the 1920s and 1930s in Europe and later on in the United States. However, as we shall see, the golden age of the principle of analyticity of logic came through an uneven route.

4.1.1 Kant

The analytic-synthetic distinction is indissolubly connected to the work of Immanuel Kant (1724–1804). This is not because he was the first to introduce it, for there are several precursors in this sense (Kant, 1997, p. 22), but rather because of the key role played by the distinction in his theoretical construction. In his *Critique of Pure Reason* (1781,1787^2), Kant seems to provide no less than four criteria for distinguishing analytic from synthetic judgements. Nevertheless, it can be shown (see Larese, 2022; De Jong, 1995) that all of them can be reduced to the following definition, which is the most fundamental criterion:

> In all judgments in which the relation of a subject to the predicate is thought (if I consider only affirmative judgments, since the application to negative ones is easy) this relation is possible in two different ways. Either the predicate B belongs to the subject A as something that is (covertly) contained in this concept A; or B lies entirely outside the concept A, though to be sure it stands in connection with it. In the first case I call the judgment **analytic**, in the second **synthetic** (Kant, 1998, A 6-7/B 10-11).

This definition soon became a touchstone for all philosophers who wished to use the analytic-synthetic distinction in their work, which gave rise to an intense debate on the details of Kant's conception. As the quotation makes clear, the containment criterion is restricted to i) true, ii) affirmative and iii) categorical (i.e., of the subject-predicate form) judgements. The latter specification has been at the core of some harsh criticism, because it implies that the analytic-synthetic distinction is not exhaustive: every non-categorical judgement is neither analytic nor synthetic (see, for example, Frege 1960, §88, pp. 99–100). But the reason why Kant did not want his distinction to be exhaustive is that he probably aimed at arguing for the syntheticity of certain judgements that in his days would have been assumed to have the subject-predicate form (see Proops, 2005)). Moreover, despite some objections raised soon after the publication of the first *Critique*, the containment criterion provides an objective and not metaphorical distinction. This is

4.1. IS LOGIC ANALYTIC?

because Kant's notion of containment is based on the Porphyrian concept hierarchies, namely a hierarchical classification of genera and species from substance in general down to individuals, which for Kant and his contemporaries, was quite a technical theory at the core of the discipline of logic (see Anderson, 2015).

The fact that logic provides the fundamental instruments that Kant employed for drawing his analytic-synthetic distinction, together with his theory that the principle of non-contradiction has the epistemological function of discerning whether judgements are analytic or not, has probably misled the great majority of Kantian readers, who hold that logic is, for Kant, analytic.[1] Nevertheless, a closer look at the texts reveals that Kant is not explicit on this point: on the contrary, he simply does not apply his analytic-synthetic distinction to logic at all. But why? The most probable reason must be found in Kant's conception of logic. For him, this discipline is formal, in that it abstracts from the content of thinking and cannot extend our knowledge of reality ((Kant, 1998, A 55/B79, A 60/B85)). Although being a science, logic as propaedeutic is not a kind of knowledge:

> [...] hence logic as a propaedeutic constitutes only the outer courtyard, as it were, to the sciences; and when it comes to information, a logic may indeed be presupposed in judging about the latter, but its acquisition must be sought in the sciences properly and objectively so called (Kant, 1998, B ix).

Nevertheless, in drawing his analytic-synthetic distinction, Kant's interests seem to stay with doctrines having a content of knowledge, such as mathematics or sciences, while logic is not considered from this perspective.

A legitimate question is now the following: is logic analytic according to Kant's definitions and independently of his reasons for leaving the matter unsolved? In attempting an analysis Kant did not want to pursue, we have that: i) no logical judgement is synthetic; ii) at least some logical truths are not analytic. As for the latter claim, recall that Kant's definition of analytic judgements via containment is restricted only to *categorical* propositions, while there are logical truths that do not have and cannot be reduced to such a form. The first class of truths that are not analytic in so far as

---

[1] See, for example, (Bolzano, 2014, §315, vol. 3, p. 161), (Hanna, 2001, p. 140), (Anderson, 2015, p. 103).

being irreducible to the form "S is P" includes all those validities turning essentially on relations. While these kinds of truths are clearly part of modern logic, traditional logic was instead intrinsically monadic in character, since it was not equipped with dedicated instruments for handling relations. The obstacle to the development of a logic of relations in the proper sense must be searched in the ontology of relations. For, since the Middle Ages, it was generally rejected that polyadic expressions of the language referred to some kind of polyadic property of the external world (see Mugnai, 2016). As a result, truths turning essentially on relations are not analytic, in so far as they cannot be reduced to categorical propositions; but it is still possible for them to be synthetic.

But the fact that there is a class of truths that belong to logic for us but not for Kant and are non-analytic does not allow us to conclude that the problem lies within the discrepancy between Kant's and contemporary definitions of logic and that every truth that Kant would have considered as logical is by default analytical. On the contrary, there is a second class of truths that do indeed belong to pure general logic even for Kant and at the same time are not reducible to categorical propositions. This is the class that includes the logical truths describing propositional inferences such as *modus ponens* or *modus tollens* (e.g., "if $A \to B$ is true and A is true, then B is true"). Such logical truths cannot be turned into categorical judgements because the former rely on the relation between judgements independently of the concepts involved. Nevertheless, inferences such as *modus ponens* and the like are for sure logical arguments: Kant includes them in the theory of inferences that he presents in his lectures on logic. As a result, *there is at least one class of logical truths that are not analytic according to Kant's definitions*.

4.1.2 Frege

It is with Gottlob Frege (1848–1925) and his book *Foundations of Arithmetic* (1884) that the relation between logic and analyticity gets stronger. In one of their rare occurrences, the terms "analytic" and "synthetic" are defined in the following way:

> The problem becomes, in fact, that of finding the proof of the proposition, and of following it up right back to the primitive truths. If,

4.1. IS LOGIC ANALYTIC?

> in carrying out this process, we come only on general logical laws and on definitions, then the truth is an analytic one, bearing in mind that we must take account also of all the propositions upon which the admissibility of any of the definitions depends. If, however, it is impossible to give the proof without making use of truths which are not of a general logical nature, but belong to the sphere of some special science, then the proposition is a synthetic one (Frege, 1960, §3, p. 4).

This definition, according to which a proposition is analytic if it can be proved with the help of general logical laws from definitions only, depends on two main premises. First, Frege's central aim throughout his work, the so-called "logicist project", is to reduce arithmetic to logic. He introduces the analytic-synthetic distinction as a counterpart of his programme: demonstrating that arithmetic can be reduced to logic amounts to showing that it is analytic. As Proust rightly observes, "the significance of analyticity is now attached to a concrete project: it must be the mortar for building up the system of arithmetic (Proust, 1989, p. 111)". Thus, the way in which Frege distinguishes between analytic and synthetic propositions might be seen as strongly influenced by his logicist program: analytic propositions are defined as those that can be proved through logical means alone, *because* Frege's aim was to show that arithmetic could be proved from logic alone.

The second premise is that Frege replaces the traditional and Kantian analysis of propositions into subject and predicate concepts with a function-argument analysis thanks to the enormous advancements in logic that he proposed in his *Begriffsschrift* (1879). According to the latter kind of analysis, propositions are analysed into a variable and a constant part. This provides a more flexible method for bringing out logically relevant similarities between sentences beyond their linguistic presentation. In this framework, categorical judgments lose the centrality they enjoyed in Kant's work, and the appeal to the notion of containment of the predicate concept into the subject concept for defining analyticities is now insufficient.

Frege's proposal disagrees with Kant's theory not only for the definition of the analytic-synthetic distinction, but also for the underlying conception of logic. According to (Frege, 1984, p. 338), logic is a body of substantive truths and not of empty schemata: logic is a proper science, which has its own objects and content. As a result, although Frege is not explicit on this point, logical propositions, in the *Foundations of Arithmetic*, are assumed to be analytic (see Larese (2019)). In particular, the laws of logic that are chosen as axioms of the system are trivially analytic, while logical theorems are analytic in that they can be proved through logical laws only. In so doing, Frege's work might be seen as an anticipation of the logical empiricists' position that takes logic to be the paradigmatic example of analyticity. Nevertheless, as it will be clarified in Sections 4.2 and 4.4, there is a fundamental difference between the two perspectives: while Frege believes that logic is not trivial, the logical empiricist movement argues that logic is devoid of any informational content.

4.1.3 Logical empiricism

The twentieth century is marked by an unprecedented interest in language as a subject matter and a radical reconception of the nature and the methods of philosophy. This revolution usually goes under the name of "linguistic turn", an expression introduced in the sixties by a former member of the Vienna Circle, Gustav Bergmann, and made popular by Rorty (1967)'s homonymous anthology. The linguistic turn, which is usually taken to coincide with the birth of the analytic philosophy, has its roots in the advances in logic (led by Frege and Russell among all) at the end of the nineteenth century. It however originated with the foundational *Tractatus logico-philosophicus* (1921), where Ludwig Wittgenstein (1889–1951) held the thesis that most philosophical problems consist in misleading features of the surface grammar of natural language and can be solved only through a logico-linguistic analysis (see Section 4.3).

These ideas were further developed by the logical empiricist movement. In the manifesto of the Vienna Circle, one of the most influential groups in this philosophical *milieu*, Rudolf Carnap, Hans Hahn and Otto Neurath write:

> The scientific world-conception knows no unconditionally valid knowledge derived from pure reason, no "synthetic judgements a priori" of the kind that lies at the basis of Kantian epistemology [...] It is precisely in the rejection of the possibility of synthetic knowledge a priori that the basic thesis of modern empiricism lies. The scientific world-conception knows only empirical statements about things of

4.1. IS LOGIC ANALYTIC?

all kinds, and analytical statements of logic and mathematics (Carnap *et al.*, 1973, p. 308).

The logical empiricist movement garnered from the *Tractatus* the so-called "verifiability criterion of meaning". According to this principle, analytic sentences, such as those of logic and mathematics, would be true in virtue of the *meaning* of their terms (and confirmed by logical analysis), while synthetic sentences, such as those of empirical sciences, would have to admit some empirical verification criteria (and confirmed by observations or sense-data). On the one hand, logical analysis is the method of clarification of philosophical problems; on the other hand, experience is the only source of knowledge.

Any sentence that could not be verified by the former or the latter of these means was deemed meaningless. A result of these assumptions is that a time-honoured discipline, such as metaphysics, is harshly criticised by logical empiricism because it relies on the ambiguity of natural language and claims that it can produce knowledge on its own sources without using any empirical material. Not only metaphysics, but the whole category of the "synthetic *a priori*", the possibility of which has been famously investigated by Kant in his *Critique of Pure Reason*, is now rejected *in toto*. In other words, according to the neo-empiricist perspective, the analytic-synthetic distinction on the one hand and the *a priori-a posteriori* distinction on the other, do now coincide, because synthetic statements are always grounded in facts and analytic statements are known *a priori*.

A consequence of this philosophical framework is that logical judgements are the paradigmatic examples of analytic judgements. The conception of logical deduction as analytic is a persistent dogma of (logical) empiricism: it caught on and became part of the logical folklore.

4.1.4 Carnap and Quine

In his paper *Two Dogmas of Empiricism* (1951), William V.O. Quine (1908–2000) famously attacked the logical empiricists' conception of semantic analyticity and their very idea that it is possible to formulate a rigorous distinction between analytic and synthetic statements. The most notable target of his criticism was Rudolf Carnap (1891–1970), a major member of the Vienna Circle, who replied to Quine in many occasions, giving rise to a debate that originated abundant philosophical literature.

Both (Quine, 1951) and (Carnap, 1952) agree that truths that are analytic by "general philosophical acclaim" fall into two classes, which can be typified by the following examples:

1. Fido is black or Fido is not black;

2. If Jack is a bachelor, then he is not married.

As (Carnap, 1952, p. 65–66) explains, both statements are true by virtue of meaning; but while in the former only the meaning of logical particles is concerned, the latter requires the analysis of the meaning of non-logical particles:

> In either case it is sufficient to understand the statement in order to establish its truth; knowledge of (extra-linguistic) facts is not involved. However, there is a difference. To ascertain the truth of (1), only the meanings of the logical particles ('is,' 'or,' 'not') are required; the meanings of the descriptive (i.e., nonlogical) words ('Fido,' 'black') are irrelevant (except that they must belong to suitable types). For (2), on the other hand, the meanings of some descriptive words are involved, viz., those of 'bachelor' and 'married'.

Quine's point in his famous (1951) paper was that "we lack a proper characterization of this second class of analytic statements" for, in his view, "the major difficulty lies not in the first class of analytic statements, the logical truths, but rather in the second class, which depends on the notion of synonymy" (Quine, 1961, pp. 22–32). In his first public response to Quine's criticism, Carnap (1952) recognized the point and suggested that analytic statements are those which can be derived from a set of appropriate sentences, which he called "meaning postulates", that define the meaning of such non-logical terms.

However, there is a growing consensus that Carnap and Quine were, to some extent, talking past each other. The main point is that while Quine's criticism focused on the possibility of defining the analytic-synthetic distinction for existing natural or scientific languages, Carnap's main interest, at least before Quine's attack, concerned giving a definition to the analytic-synthetic distinction for *specific constructed artificial languages* (Leitgeb & Carus, 2024). In other words, although Carnap's account of analyticity underwent several significant changes, all of his formulations are characterised

4.1. IS LOGIC ANALYTIC?

by the idea that the notion of analytic truths is relative to a certain language or, to be more precise, that analytic truths define what makes something into a language. In this restricted domain, Quine's criticism does not seem to have a bite.

Four decades later, while his reservations over the notion of analyticity remained the "the same as ever", Quine clarified that they concerned only "the tracing of any demarcation, even a vague and approximate one, across the domain of sentences in general" (Quine, 1991, p. 270). But the impossibility of tracing a demarcation does not exclude that there may be undebatable cases of analytic sentences. Indeed, "It is intelligible and often useful in discussion to point out that some disagreement is purely a matter of words rather than of fact" (Quine, 1991, p. 270). The so-called "logical laws" are the most natural candidates for such paradigmatic examples of analytic sentences: it seems almost uncontroversial that a disagreement about a logical truth can always be reduced to a disagreement about the meaning of some logical word that occurs in it.

In fact, in *The Roots of Reference* (1974) Quine had already suggested that, in order to fit the undisputed cases of analytic sentences, one may provide a rough theoretical definition of analyticity by saying that (i) a sentence is analytic for the native speakers of a language if they learn its truth in the very process of learning how to use the words that occur in it, and (ii) "recondite" sentences should still count as analytic if they can be obtained by "a chain of inferences each of which individually is assured by the learning of the words" (Quine, 1974, pp. 22–32). In this perspective, logical truths may qualify as analytic in the traditional sense, although the very existence of enduring disagreement on some logical laws — e.g. on the law of excluded middle on the part of intuitionists — may suggest that such laws are not similarly bound up with the learning of the logical words and "should perhaps be seen as synthetic" (Quine, 1974, pp. 22–32).

In his latest work, Quine appears to leave aside this idea that some logical laws may be synthetic. For example, in his *Two Dogmas in Retrospect* (1991), he argues that by the above criterion "all logical truths [...] — that is, the logic of truth functions, quantification, and identity — would then perhaps qualify as analytic, in view of Gödel's completeness proof" (Quine, 1991, p. 270) and later on, in a 1993 interview, seems to abandon any hesitation and make his position crystal-clear:

> Yes so, on this score I think of the truths of logic as analytic in the traditional sense of the word, that is to say true by virtue of the meaning of the words. Or as I would prefer to put it: they are learned or can be learned in the process of learning to use the words themselves, and involve nothing more (Bergström & Føllesdag, 1994, p. 199).[2]

4.2 Is logic tautological?

As we have seen in the previous section, the characteristic of being "analytical", when used to qualify logical reasoning, is typically interpreted along the following lines:

> *Semantic or meaning-theoretical analyticity.* An inference is analytical in the semantic or meaning-theoretical sense if, and only if, whenever the premises are true, the conclusion must also be true by virtue only of the accepted meaning of the logical operators. This sense of "analytical" usually comes with a theory of how the meaning of the logical operators is specified.

Given the meaning of the logical operators, the conclusion must necessarily be true whenever the premises are true. In most logical systems valid deductive inferences are defined in such a way that they turn out to be analytic in this sense. For example, in classical propositional logic, valid inferences are exactly those that are faithful to, and licensed by, the truth-tables for the logical operators, which are usually taken as definitions of their meaning.

This semantic sense of the word "analytic", when applied to an inference, is usually associated with another sense, that we may call its *informational sense*, according to which an analytic inference is one that is "tautological", i.e. does not increase information. This sense can be made more explicit as follows:

> *Informational analyticity or tautologicity.* An inference is analytical in the informational sense or tautological if, and only if, whenever we possess the information that the premises are true, we thereby possess the information that the conclusion is true. This sense of "analytical" usually comes with an explanation of the notion of "possessing the information" that a certain sentence is true or false.

---
[2]Quoted in Decock (2006).

4.2.1 The scandal of deduction

One answer comes from the "paradox of analysis", which is a time-honoured paradox in the history of philosophy and can be described in the following terms:

> Let us call what is to be analyzed as the analysandum, and let us call that which does the analyzing the analysans. The analysis then states an appropriate relation of equivalence between the analysandum and the analysans. And the paradox of analysis is to the effect that, if the verbal expression representing the analysandum has the same meaning as the verbal expression representing the analysans, the analysis states a bare identity and is trivial; but if the two verbal expressions do not have the same meaning, the analysis is incorrect (Langford, 1992, p. 323).

This paradox points at the incompatibility of two desirable properties for analysis: correctness and informativity. Now, assuming that logic is analytic amounts to say that logic is the result of some kind of analysis. In particular, following the semantical perspective of logical empiricism, the truth of a logical inference results from the analysis of the meaning of the logical operators involved. But here the paradox of analysis does show all of its force: if logic is analytic, then (or so it seems) either it is incorrect or it must be tautologous, namely trivial and non-informative. Of course, the first option is not viable. As a result, if in an inference the conclusion follows from the premises only by virtue of the meaning of the logical words occurring in the sentences (i.e., the inference is analytic in the semantic sense), then the conclusion can be "seen" as true, whenever its premises are true, and the inference carries no new information (i.e., the inference is tautological). This means that the principle of analyticity of logic, together with the paradox of analysis, implies that logic is tautologous and trivial: since it is correct, logic cannot yield new information.

As pointed out by Cohen and Nagel, however, this trivialization of logic sounds paradoxical:

> If in an inference the conclusion is not contained in the premises, it cannot be valid; and if the conclusion is not different from the

premises, it is useless; but the conclusion cannot be contained in the premises and also possess novelty; hence inferences cannot be both valid and useful (Cohen & Nagel, 1934, p. 173).

However, the logical empiricist movement accepts the paradox and its seemly inescapable consequence that logic and mathematics are tautologous. In the manifesto of the Vienna Circle, Carnap, Hahn and Neurath write:

> Logical investigation [...] leads to the result that all thought and inference consists of nothing but a transition from statements to other statements that contain nothing that was not already in the former (tautological transformation) [...] The conception of mathematics as tautological in character, which is based on the investigation of Russell and Wittgenstein, is also held by the Vienna Circle (Carnap et al., 1973, p. 308 and p. 311).

This thesis held by the Vienna Circle is influenced by the analysis of logical truth proposed by Wittgenstein in his *Tractatus Logico-Philosophicus* (1921). While Frege treats logical truths as universal laws applying to any statement, Wittgenstein believes that the laws of logic are tautologies, which, in themselves, do not say anything. According to him, tautologies, which cannot be false, do not tell us how the world in fact is: "tautologies and contradictions show that they say nothing. A tautology has no truth-conditions, since it is unconditionally true [...] (For example, I know nothing about the weather when I know that it is either raining or not raining.)" (Wittgenstein, 2001, 4.461). A result of this standpoint is that tautologies, as well as contradictions, lack sense.

But the traditional tenet that logic is tautologous raises several conceptual difficulties. First and foremost, the tenet is highly counterintuitive and can hardly be accepted. The conclusion obtained through a long deductive chain might appear as an actual novelty with respect to its premises, and the recognition that a particularly complex sentence is a tautology might appear as a true discovery. In his work, Jaakko Hintikka recognises the point and describes this situation as a true "scandal of deduction":

> C.D. Broad has called the unsolved problems concerning induction a scandal of philosophy. It seems to me that in addition to this scandal of induction there is an equally disquieting scandal of deduction. Its urgency can be brought home to each of us by any clever freshman

4.2. IS LOGIC TAUTOLOGICAL?

who asks, upon being told that deductive reasoning is "tautological" or "analytical" and that logical truths have no "empirical content" and cannot be used to make "factual assertions": in what other sense, then, does deductive reasoning give us new information? Is it not perfectly obvious there is some such sense, for what point would there otherwise be to logic and mathematics? (Hintikka, 1973, p. 222).

The scandal of deduction is not confined to classical propositional logic, but carries over to the widely studied logics of knowledge (epistemic logic) and belief (doxastic logic), as well as to the more recent attempts to axiomatize the "logic of being informed" (information logic).[3] If an agent a knows (or believes, or is informed) that a sentence A is true, and B is a logical consequence of A (thus, according to the received view, B is a trivial and tautological consequence of A), then a is supposed to know (or believe, or be informed) also that B is true. This is often described as paradoxical and labelled as "the problem of logical omniscience". Let $\Box_a$ express any of the propositional attitudes at issue, relative to the agent a. Then, the "logical omniscience" assumption can be expressed by saying that, for any finite set Γ of sentences,

(4.1) $\quad\quad$ if $\Box_a A$ for all $A \in \Gamma$ and $\Gamma \vdash B$, then $\Box_a B$,

where $\vdash$ stands for the relation of logical consequence. Observe that, letting $\Gamma = \emptyset$, it immediately follows from (4.1) that any rational agent a is supposed to be aware of the truth of all classical tautologies, that is, of all the sentences of a standard logical language that are "consequences of the empty set of assumptions". In most axiomatic systems of epistemic, doxastic and information logic, assumption (4.1) emerges from the combined effect of the "distribution axiom", namely,

(K) $\Box_a(A \to B) \to (\Box_a A \to \Box_a B)$

and the "necessitation rule":

(N) if $\vdash A$, then $\vdash \Box_a A$.

On the other hand, despite its paradoxical flavour, (4.1) seems an inescapable consequence of the standard Kripke-style semantical characterization of the

---
[3]For a survey on epistemic and doxastic logic see Halpern (1995); Meyer (2003). For information logic, or "the logic of being informed", see Floridi (2006); Primiero (2009).

logics under consideration. The latter is carried out in terms of structures of the form $(S, \tau, R_1, \ldots, R_n)$, where S is a set of possible worlds, τ is a function that associates with each possible world s an assignment $\tau(s)$ of one of the two truth values (0 and 1) to *each* atomic sentence of the language, and each R_a is the "accessibility" relation for the agent a. Intuitively, if s is the actual world and $sR_a t$, then t is a world that a would regard as a "possible" alternative to the actual one, i.e., compatible with what a knows (or believes, or is informed of). Then, the truth of complex sentences is defined, starting from the initial assignment τ, via a forcing relation $\vDash$. This incorporates the usual semantics of classical propositional logic and defines the truth of $\Box_a A$ as "A is true in all the worlds that a regards as possible". In this framework, given that the notion of truth in a possible world is an extension to the modal language of the classical truth-conditional semantics for the standard logical operators, (4.1) appears to be both compelling and, at the same time, counter-intuitive.

Now, under this reading of the consequence relation $\vdash$, which is based on classical propositional logic, (4.1) may perhaps be satisfied by an "idealized reasoner", in some sense to be made more precise, but, as we have claimed above, it is not satisfied, and is not likely to ever be satisfiable, in practice. So, the clash between (4.1) and the classical notion of logical consequence which arises in any real application context, may only be solved either by waiving the assumption stated in (4.1), or by waiving the consequence relation of classical logic in favour of a weaker one with respect to which it may be safely assumed that the modality $\Box_a$ is closed under logical consequence for any realistic agent.

Both options have been discussed in the literature.[4] Observe that, according to the latter, the problem of logical omniscience does not lie in

---

[4]See Meyer (2003, Section 4), Halpern (1995, Section 4), Sim (1997) and Égré (2020) for a survey and proper references. See also: Parikh (2008) for an interesting third view that draws on the tradition of subjective probability, Artemov & Kuznets (2009) for an approach based on proof size, and Bjerring & Skipper (2019) and Hawke *et al.* (2020b) for a dynamic approach. A general semantic framework in which several different approaches can be usefully expressed is that based on "awareness structures", which draws on the distinction between "explicit" and "implicit" knowledge, to the effect that an agent may implicitly know that a sentence is a logical consequence of a set of assumptions, without being *aware* of it. See Sillari (2008a,b) for an insightful discussion of this framework and proper references to the literature.

4.2. IS LOGIC TAUTOLOGICAL?

assumption (4.1) in itself, but rather in the standard (classical) characterization of logical consequence for a propositional language that is built in the possible-world semantics originally put forward by Jaakko Hintikka as a foundational framework for the investigation of epistemic and doxastic logic. Frisch (1987) and Levesque (1988) were among the first authors to explore this route and argue for a notion of "limited inference" based on "a less idealized view of logic, one that takes very seriously the idea that certain computational tasks are relatively easy, and others more difficult (Levesque, 1988, p. 355)". Another related proposal can be found in (Fagin *et al.*, 1995a), where the authors suggest to replace classical logic with a non-standard one, deeply rooted in relevance logic and called NPL (for "Nonstandard Propositional Logic"), to mitigate the problem of logical omniscience.

4.2.2 Semantic information in propositional logic

At the half of the 20th century Bar-Hillel and Carnap's theory of "semantic information" provided what is, to date, the strongest theoretical justification for the thesis that deductive reasoning is "tautological". Although their effort was clearly inspired by the rising enthusiasm for Shannon and Weaver's new Theory of Information (Shannon & Weaver, 1949), their starting point was their dissatisfaction with the nonchalant tendency of fellows scientists to apply its concepts and results well beyond the "warranted areas". Shannon and Weaver's central problem was only how uninterpreted data can be efficiently encoded and transmitted. So the idea of applying their theory to contexts in which the *interpretation* of data plays an essential rôle was a major source of confusion and misunderstandings:

> The Mathematical Theory of Communication, often referred to also as "Theory of (Transmission of) Information", as practised nowadays, is not interested in the content of the symbols whose information it measures. The measures, as defined, for instance, by Shannon, have nothing to do with what these symbols symbolise, but only with the frequency of their occurrence. [...] This deliberate restriction of the scope of the Statistical Communication Theory was of great heuristic value and enabled this theory to reach important results in a short time. Unfortunately, however, it often turned out that impatient scientists in various fields applied the terminology and the theorems of

> Communication Theory to fields in which the term "information" was used, pre-systematically, in a semantic sense, that is, one involving contents or designata of symbols, or even in a pragmatic sense, that is, one involving the users of these symbols. (Bar-Hillel & Carnap, 1953, p. 147)

By way of contrast, they put forward a Theory of *Semantic* Information, in which "the contents of symbols" were "decisively involved in the definition of the basic concepts" and "an application of these concepts and of the theorems concerning them to fields involving semantics thereby warranted" (Bar-Hillel & Carnap, 1953, p. 148). The basic idea is simple and can be briefly explained as follows.

Suppose we are interested in the weather forecast for tomorrow and that we focus only on the possible truth values of the two sentences "tomorrow will rain" (R) and "tomorrow will be windy" (W). Then, there are four possible relevant states of the world, described by the following conjunctions:

$$R \wedge W \quad R \wedge \neg W \quad \neg R \wedge W \quad \neg R \wedge \neg W.$$

Now, the sentence "tomorrow it will rain and will be windy" is intuitively more informative than the sentence "tomorrow it will rain". We can explain this by noticing that it excludes more possibilities, i.e, more possible (relevant) states of the world. On the other hand, the sentence "tomorrow it will rain or will not rain" conveys no information, since it does not exclude any possible state. So, it seems natural to identify the information conveyed by a sentence with the set of all "possible worlds" that are excluded by it, and to assume that its measure should be somehow related to the size of this set.

The same basic idea, identifying the information carried by a sentence with the set of the possible states that it excludes, had already made its appearance in Popper's *Logic of Scientific Discovery* (1934), where it played a crucial rôle in defining the "empirical content" of a theory and in supporting Popper's central claim, namely that the most interesting scientific theories are those that are highly falsifiable, while unfalsifiable theories are devoid of any empirical content:

> The amount of positive information about the world which is conveyed by a scientific statement is the greater the more likely it is to clash, because of its logical character, with possible singular statements. (Not for nothing do we call the laws of nature "laws": the more they prohibit the more they say.) (Popper, 1959, p. 19).

4.2. IS LOGIC TAUTOLOGICAL?

> [...]
> It might then be said, further, that if the class of potential falsifiers of one theory is "larger" than that of another, there will be more opportunities for the first theory to be refuted by experience; thus compared with the second theory, the first theory may be said to be "falsifiable in a higher degree". This also means that the first theory says more about the world of experience than the second theory, for it rules out a larger class of basic statements. [...] Thus it can be said that the amount of empirical information conveyed by a theory, or its empirical content, increases with its degree of falsifiability (Popper, 1959, p. 96).

An inevitable consequence of the theory of semantic information is that *all* logical truths are equally uninformative (they exclude no possible world), which justifies their being labelled as "tautologies". But in classical logic a sentence φ is deducible from a finite set of premises $\psi_1, \ldots, \psi_n$ if and only if the conditional $(\psi_1 \wedge \ldots \wedge \psi_n) \to \varphi$ is a tautology. Accordingly, since tautologies carry no information at all, no logical inference can yield an increase of information. Therefore, if we identify the semantic information carried by a sentence with the set of all possible worlds it excludes, we must also accept the inevitable consequence that, in any valid deduction, the information carried by the conclusion is contained in the information carried by the (conjunction of) the premises.

4.2.3 The BHC paradox

Another straightforward consequence of Bar-Hillel and Carnap's notion of semantic information is that contradictions, like "tomorrow it will rain and it will not rain", carry the maximum amount of information, since they exclude all possible states. This is the so-called "Bar-Hillel and Carnap (BHC) paradox": while tautologies are devoid of any informational content, because they are true in every possible state, then contradictions, like "tomorrow it will rain and it will not rain", carry instead the maximum amount of information, since they exclude all possible states. Bar-Hillel and Carnap were well aware that their theory of semantic information sounded counter-intuitive in connection with contradictory (sets of) sentences, as shown by the near-apologetic remark they included in their report:

> It might perhaps, at first, seem strange that a self-contradictory sen-

tence, hence one which no ideal receiver would accept, is regarded as carrying with it the most inclusive information. It should, however, be emphasized that semantic information is here not meant as implying truth. A false sentence which happens to say much is thereby highly informative in our sense. Whether the information it carries is true or false, scientifically valuable or not, and so forth, does not concern us. A self-contradictory sentence asserts too much; it is too informative to be true (Bar-Hillel & Carnap, 1964, p. 229).

Popper had also realized that his closely related notion of empirical content worked reasonably well only for *consistent* theories. For, *all* basic statements are potential falsifiers of all inconsistent theories, which would therefore, without this requirement, turn out to be the most scientific of all. So, for him, "the requirement of consistency plays a special role among the various requirements which a theoretical system, or an axiomatic system, must satisfy" and "can be regarded as the first of the requirements to be satisfied by every theoretical system, be it empirical or non-empirical (Popper, 1959, p. 72)". So, "whilst tautologies, purely existential statements and other unfalsifiable statements assert, as it were, *too little* about the class of possible basic statements, self-contradictory statements assert *too much*. From a self-contradictory statement, any statement whatsoever can be validly deduced (Popper, 1959, p. 71)". In fact, what Popper claimed was that the information content of inconsistent theories is null, and so his definition of information content as monotonically related to the set of potential falsifiers was intended only for consistent ones:

> But the importance of the requirement of consistency will be appreciated if one realizes that a self-contradictory system is uninformative. It is so because any conclusion we please can be derived from it. Thus no statement is singled out, either as incompatible or as derivable, since all are derivable. A consistent system, on the other hand, divides the set of all possible statements into two: those which it contradicts and those with which it is compatible. (Among the latter are the conclusions which can be derived from it.) This is why consistency is the most general requirement for a system, whether empirical or non-empirical, if it is to be of any use at all (Popper, 1959, p. 72).

To summarize, the received view that logic is analytical and tautological has well-known issues and its consequences clash with the layman's intu-

4.3 Attempts to explain the scandals

4.3.1 Logical empiricism

ition. This raises an obvious question: how do philosophers, who hold this view, manage to cope with its anomalies?

One answer comes from the logical empiricist *milieu* and has a psychologistic flavour. In his pamphlet *Logic, Mathematics and Knowledge of Nature* written in 1933, Hans Hahn (1879–1934) recognizes that there are two sources of knowledge, experience and thinking, and reconstructs the controversy in the history of philosophy about the relationship between the two. While rationalism failed because its fruits lacked nourishing value, the earlier empiricists committed the error of interpreting the propositions of logic and mathematics as mere facts of experience. Thus, according to the neo-empiricist mathematician Hahn, a different conception of logic and mathematics is needed. Hahn argues that "logic does not by any means treat of the totality of things, it does not treat of objects at all but *only of our way of speaking about objects* (Hahn, 1959, p. 152)": in other words, logic is generated by language. Moreover, influenced by Wittgenstein, he holds that the statements of logic, that express the way in which the rules that govern the application of words to facts depend upon each other, are tautologies: "they say nothing about objects and are for this very reason certain, universally valid, irrefutable by observation (Hahn, 1959, p. 153)". In conveying this interpretation of logic, Hahn is thus choosing one horn of the paradox of analysis, namely, that logic is correct and uninformative. But, at the same time, he offers an answer to the question as to what purpose logic serves:

> Thus logical propositions, though being purely tautologous, and logical deductions, though being nothing but tautological transformations, have significance for us because we are not omniscient. Our language is so constituted that in asserting such and such propositions we implicitly assert such and such other propositions – but we do not see immediately all that we have implicitly asserted in this manner. It is only logical deduction that makes us conscious of it. [...] The propositions of mathematics are of exactly the same kind as the propositions of logic (Hahn, 1959, pp. 157–158).

According to Hahn, logic and mathematics are not objectively new or informative. They are instruments that simply compensate for our limitations and our inability to see immediately the consequences of what we know: "an omniscient being has no need for logic and mathematics (Hahn, 1959, p. 159)". Using the words that Hintikka chooses to criticize this position (see Section 4.4), we could say that "all that is involved is merely psychological conditioning, some sort of intellectual psychoanalysis, calculated to bring us to see better and without inhibitions what objectively speaking is already before our eyes (Hintikka, 1973, p. 223)". Logic and mathematics make us conscious of the consequences of our premises that we are not intelligent enough to recognize by mere inspection. This idea, which will be distinctive of the analytic tradition, is a reformulation of Wittgenstein's theory that philosophy is the activity of clarification of thought:

> Philosophy aims at the logical clarification of thoughts. Philosophy is not a body of doctrine but an activity. A philosophical work consists essentially of elucidations. Philosophy does not result in 'philosophical propositions', but rather in the clarification of propositions. Without philosophy thoughts are, as it were, cloudy and indistinct: its task is to make them clear and to give them sharp boundaries (Wittgenstein, 2001, 4.112).

and is unconditionally accepted by the logical empiricist movement:

> Clarification of the traditional philosophical problems leads us partly to unmask them as pseudo-problems, and partly to transform them into empirical problems and thereby subject them to the judgment of experimental science. The task of philosophical work lies in the clarification of problems and assertions, not in the propounding of special philosophical pronouncements. The method of this clarification is that of *logical analysis* (Carnap et al., 1973, p. 306).

A similar position is put forward by the mathematician and philosopher Carl Gustav Hempel (1905–1997). In his article *On the Nature of Mathematical Truth* first published in 1945, he holds that the validity of mathematics depends neither on its alleged self-evident character, nor on any empirical basis, but derives by virtue of definitions and stipulations which determine the meaning of the terms. Mathematical and logical statements

4.3. ATTEMPTS TO EXPLAIN THE SCANDALS

are called analytic and their truth is independent of any experiential evidence. However, the price paid for the theoretical certainty of these disciplines is very high: they convey no factual information. In another text of the same year, *Geometry and Empirical Science*, Hempel suggests that logic might be psychologically useful along the same lines indicated by Hahn:

> Logical deduction – which is the one and only method of mathematical proof – is a technique of conceptual analysis: it discloses what assertions are concealed in a given set of premises, and it makes us realize to what we committed ourselves in accepting those premises; but none of the results obtained by this technique ever goes by one iota beyond the information already contained in the initial assumptions [...] a mathematical theorem, such as the Pythagorean theorem in geometry, asserts nothing that is *objectively* or *theoretically new* as compared with the postulates from which it is derived, although its content may well be *psychologically new* in the sense that we were not aware of its being implicitly contained in the postulates (Hempel, 2001, pp. 20–21).

A slightly different perspective on the same issue is given by Alfred Jules Ayer (1910–1989), who dedicates the fourth chapter of his book entitled *Language, Truth and Logic* written in 1936 to answer to the traditional objection that it is impossible on empiricist principles to account for our knowledge of necessary truths. After having recognized the problem posed by the paradox of analysis, he argues that, on the one hand, analytic propositions do not give us new knowledge, for they are devoid of factual content and consequently they say nothing. Nevertheless, he maintains, on the other hand, that there is a sense in which analytic propositions might add something to our knowledge: "although they give us no information about any empirical situation, they do enlighten us by illustrating the way in which we use certain symbols (Ayer, 1958, p. 80)". Ayer's idea is that logical deduction calls attention to the implications of a certain linguistic usage, such as the convention which governs our employment of the connectives, of which we might otherwise not be conscious. Hempel and Hahn found that the power of logic to surprise us depended on the recognition of the consequences that could not be immediately grasped because of human limitations; Ayer is more specific and maintains that logical deduction sheds new light in particular on the functioning of certain linguistic items and, ultimately, of our logical systems. This difference notwithstanding, Ayer's solution to the paradox of analysis is as psychologistic in character as the previous ones, because the British philosopher also holds that the novelty of the result of a logical deduction depends on the limitation of our reason:

> A being whose intellect was infinitely powerful would take no interest in logic and mathematics. For he would be able to see at a glance everything that his definitions implied, and, accordingly, could never learn anything from logical inference which he was not fully conscious of already. But our intellects are not of this order. It is only a minute proportion of the consequences of our definitions that we are able to detect at a glance (Ayer, 1958, pp. 85–86).

Ayer's thesis that logical deduction provides us with new information concerning our logical system does not, *by itself*, imply that linguistic and conceptual information cannot be perfectly objective and non-psychological. The viability of this theoretical option is shown by the work of Hintikka, which will be discussed in the next section. However, the psychologistic flavor of Hahn, Hempel and Ayer's solution is not completely satisfying. The obstacle that prevents us from deriving all the consequences of what we know, cannot be a subjective incapability of the individuals, but rather seems to be an objective barrier: the (probable) intractability of classical propositional logic.

4.3.2 Wittgenstein

A different explanation of the intuitive idea that logical deduction is fruitful, which is consistent with the trivialization of logic, comes from the ideal of the logically perfect language and its relation to the myth of instant rationality. In the *Introduction* to the *Grundlagen*, Frege radicalizes his representationalist view of thinking, according to which linguistic items are essential mediators of our grasping thoughts, and holds the following thesis:

> It is possible, of course, to operate with figures mechanically, just as it is possible to speak like a parrot: but that hardly deserves the name of thought. It only becomes possible at all after the mathematical notation has, as a result of genuine thought, been so developed that it does the thinking for us, so to speak (Frege, 1960, p. iv).

4.3. ATTEMPTS TO EXPLAIN THE SCANDALS

In a suitable notation, all logical relations become visible and thinking turns out to be superfluous. Frege's standpoint is clearly influenced by the Leibnizian tradition, to whom he explicitly refers. In the preface of his *Begriffsschrift*, he points out that Leibniz recognized the advantages of an adequate system of notation and holds that his own ideography is a characterization of Leibniz's "universal characteristic" and of "a *calculus philosophicus* or *ratiocinator*".

There are manifold Leibnizian echoes also in Wittgenstein's thought. In particular, the author of the *Tractatus Logico-Philosophicus* deals with the problem of an adequate logical notation that he characterizes in terms of the perfect coincidence of the grammatical and the logical structures of a sentence. Each sentence expressed through an adequate notation shall immediately show its sense, which is given by the conditions in which it is true or false. As a result of the use of a perfect language, every tautology shall be recognized immediately as a sentence that is true no matter the interpretation, and the validity of every inference shall be clear at first sight. With an adequate notation, logical deduction shall be completely superfluous, for a mere inspection of even complex sentences shall be sufficient to grasp their sense. As (Wittgenstein, 2001) puts it:

> 5.13 When the truth of one proposition follows from the truth of others, we can see this from the structure of the propositions.
>
> 6.122 [...] in a suitable notation we can in fact recognize the formal properties of propositions by mere inspection of the propositions themselves.
>
> 6.127 [...] Every tautology itself shows that it is a tautology.

But does anything like an adequate logical notation exist according to Wittgenstein? He rejects Frege and Russell's proof system as an example of the adequate logical notation (see Carapezza & D'Agostino, 2010). First of all, the criterion according to which some tautologies are chosen to play the role of axioms, namely, self-evidence, is shared by every single tautology and is thus useless. Moreover, not only the same proposition, that for the Austrian philosopher is identifiable with the possibility of its being true or false, can be expressed through different signs, due to the inter-definability of logical connectives, but also logical derivability itself might turn into an extremely complex task. As a result, according to Wittgenstein, Frege and Russell's proof system preclude the immediate visibility of propositions.

But Wittgenstein's reflection on the adequate notation is also endowed with a *pars construens*. He considers "truth tables", introduced in the *Tractatus* for the first time (Wittgenstein, 2001, 4.31 and 4.442), to be an adequate way to express propositions. The *Tractatus* distinguishes between elementary and complex propositions, the latter being made up by the former: while the truth of an elementary proposition depends on the existence of certain facts about the world, the truth value of a complex proposition depends on the elementary constituents of which it is constructed. The truth table of a complex proposition not only shows in an explicit way the truth conditions of that proposition, but it is in itself a propositional sign, viz. it serves as a proposition.

In what way does the myth of the perfect language offer a solution to the paradox of analysis? As we have seen in Section 4.2, Wittgenstein, as well as logical positivists, holds that logic is both analytic and tautological. However, while the psychologistic perspective *à la* Hempel, although denying that deductive reasoning has objective utility, accepts that the conclusion of a valid inference might be psychologically new with respect to its premises, Wittgenstein rejects the usefulness of logical deduction *tout court*. According to the insights conveyed in the *Tractatus*, once propositions are expressed through an adequate notation, i.e., in terms of their truth tables, logical deduction shall be replaced by the mere inspection of the propositions. Wittgenstein has thus chosen one horn of the dilemma, namely, analysis is not informative. At the same time, he offers an explanation of the apparent novelty of the conclusion of an inference with respect to its premises in terms of the inadequacy of our logical language.

However, Wittgenstein's position has two main drawbacks. On the one side, the solution proposed in the *Tractatus* is restricted to propositional logic and, crucially, Church-Turing's undecidability theorem (1936) excludes the possibility of finding a similar perfect notation for first-order logic. On the other side, there is probably no feasible translation from the ordinary language to the adequate notation, for its existence would also imply the existence of an efficient deterministic algorithm to solve the tautology problem. But unfortunately, the currently accepted conjecture in theoretical computer science that $P \neq NP$ excludes the tractability of the tautology problem.

4.4 Attempts to reject the scandal

The previous section discussed two attempts to explain the scandal of deduction, while assuming that logic is analytic and tautological. We now review two attempts to reject the scandal in order to support the idea that logical deduction is fruitful, and therefore not tautological.

4.4.1 Frege

Frege offers an original solution to the problem posed by the paradox of analysis. In the *Grundlagen der Arithmetik*, he maintains that "propositions which extend our knowledge can have analytic judgements for their content (Frege, 1960, §91, p. 104)" and, as a special case of this fact, he holds the thesis that logic is, at the same time, analytic and informative.

His starting point seems to be, once more, his logicist position that arithmetic can be reduced to logic. According to the author of the *Grundlagen*, if his logicist bet proved to be winning, "the truths of arithmetic would then be related to those of logic in much the same way as the theorems of geometry to the axioms (Frege, 1960, §17, p. 24)". In other words, the relationship between logic and arithmetic would be so close that the former would contain, albeit in a concentrated format, all the theorems of the latter discipline. At the same time, Frege assumes as a matter of fact that arithmetic cannot be charged as being sterile: the results of this discipline are so evident and exceptional that nobody could seriously maintain that arithmetical propositions are uninformative or tautologous. The natural result of combining these two premises is that the fruitfulness of arithmetic expands to cover logic, so that also the latter discipline must be equally seen as capable of extending our knowledge:

> Can the great tree of the science of number as we know it, towering, spreading, and still continually growing, have its roots in bare identities? And how do the empty forms of logic come to disgorge so rich a content? (Frege, 1960, §16, p. 22).

> [...] the prodigious development of arithmetical studies, with their multitudinous applications, will suffice to put an end to the widespread contempt for analytic judgments and to the legend of the sterility of pure logic (Frege, 1960, §17, p. 24).

162 CHAPTER 4. PROBLEMS IN THE PHILOSOPHY OF LOGIC

However, the connection between logic and arithmetic cannot be considered as an explanation in itself and a way out of the paradox of analysis. How does Frege manage to keep analyticity and informativeness together? The answer can be found in the notion of definition understood as a kind of concept formation through the process of analysis, that Frege proposed. In the unpublished text *Boole's Logical Calculus and the Concept-Script*, Frege criticizes Boole's concept formation with the following words:

> In this sort of concept formation, one must, then, assume as given a system of concepts, or speaking metaphorically, a network of lines. These really already contain the new concepts: all one has to do is to use the lines already there to demarcate complete surface areas in a new way. It is the fact that attention is principally given to this sort of formation of new concepts from old ones, while other more fruitful ones are neglected which surely is responsible for the impression one easily gets in logic that for all our to-ing and fro-ing we never leave the same spot (Frege, 1979, p. 34).

Frege explains that, according to Boole's perspective, the definition of *homo* in terms of *animal rationale* corresponds to logical multiplication and illustrates that the extension of *homo* is the intersection of two circles, the former representing the extension of the concept *animal* and the latter of *rationale*. Moreover, he notices that concepts can be formed in Boolean logic not only through multiplication, but also through addition. This is the case, for example, of the definition of "capital offence" as "murder or the attempted murder of the Kaiser or of the rules of one's own *Land* or of a German prince in his own *Land*". The extension of the concept "capital offence" is given as the union of two circles, the former representing the extension of "murder" and the latter of the second disjunct of the definition above. Addition and multiplication are the familiar ways of forming concepts but, as the quotation makes clear, they are characterized by the same method of using old lines to demarcate new surfaces. This kind of concept-formation is, in Frege's eyes, responsible for the idea that logic is sterile.

On the contrary, Frege's conception of analysis not only yields a plurality of results, due to the possibility of choosing in different ways which parts of a judgement to consider as variable and which to consider as constant. It also provides a fruitful method of concept-formation. One of (Frege, 1979, pp. 16–17)'s examples considers the equation $2^4 = 16$. This judgement

4.4. ATTEMPTS TO REJECT THE SCANDAL

can be decomposed in several ways. First, if we consider number 2 to be variable, which may be indicated as $x^4 = 16$, we obtain the concept "4th root of 16". Second, if we consider number 4 to be variable, so that we may write $2^x = 16$, we obtain the concept "logarithm of 16 to the base 2". Third, if we consider both 2 and 16 to be replaceable and we indicate this by the expression $x^4 = y$, we obtain the relation of a number to its 4th power. The point is that the concepts that have been formed in this way are actually new and fruitful:

> There's no question here of using the boundary lines of concepts we already have to form the boundaries of new ones. Rather, totally new boundary lines are drawn by such definitions — and these are the scientifically fruitful ones. Here too, we use old concepts to construct new ones, but in so doing we combine the old ones together in a variety of ways by means of the sign for generality, negation and the conditional (Frege, 1979, p. 34).

In modern terms, we could say that the fruitfulness of the process described by Frege consists in the individuation of a quantificational structure in logically unstructured judgements or, in other words, in the recognition of certain patterns, such as predicates and relations, in given propositions. The problem is to establish which is the pattern, among the many options available in every proposition, that better suits the demonstration. Moreover, it is the work of extracting these structures from judgements that yields an extension of knowledge.

In the *Grundlagen*, Frege goes one step further, for he says that it is precisely due to this new concept-formation producing fruitful definitions that analytic judgements can extend our knowledge:

> What we shall be able to infer from it [i.e., from the more fruitful type of definition], cannot be expected in advance; here, we are not simply taking out of the box again what we have just put into it. The conclusions we draw from it extend our knowledge, and ought therefore, on Kant's view, to be regarded as synthetic; and yet they can be proved by purely logical means, and are thus analytic. The truth is that they are contained in the definitions, but as plants are contained in their seeds, not as beams are contained in a house (Frege, 1960, §88, pp. 100–101).

Here, Frege is saying that conclusions that can be drawn from fruitful definitions extend our knowledge exactly because of the fertility of those definitions and despite the fact that conclusions can be proved by purely logical means alone. In other words, in this excerpt Frege maintains that the fruitfulness of definitions makes deductions that take the steps from them informative, and this happens even in the case in which deductions employ only general logical laws and, thus, produce analytic conclusions.

Moreover, he is explaining, albeit metaphorically, the link between analyticity and informativeness. On the one hand, deductions are knowledge-extending processes because conclusions are contained in the premises, i.e. the initial definitions, only *in posse*, but not *in esse*. It is in the transition from potentiality to actuality, viz. from seeds to plants, that the informativeness of deductive reasoning finds its place. This is a resource-consuming process: as (Dummett, 1991a, p. 42) puts it, this is not a mechanical procedure, but it has a rather creative component due to the requirement of pattern recognition. On the other hand, however, the conclusion of a deductive process is analytic: there is no need of any other tool than the logical ones to prove it, since the conclusion is, although only potentially, already contained in the given definitions. Informativeness and analyticity are thus two unavoidable consequences of the way Frege characterizes the notion of deduction and its dependence on fruitful definitions. This is well explained by Dummett in the following terms:

> Since it has this creative component, a knowledge of the premises of an inferential step does not entail a knowledge of the conclusion, even when we attend to them simultaneously; and so deduction can yield new knowledge. Since the relevant patterns need to be discerned, such reasoning is fruitful; but, since they are there to be discerned, its validity is not called in question (Dummett, 1991a, p. 42).

The solution to the paradox of analysis that Frege offers at the beginning of the 1880s is fascinating. It has the merit of breaking the connection between tautologicity and analyticity. The basic idea is that the recognition of a certain pattern and the extraction of a quantificational structure from a given judgement is, by itself, a creative process. Logical deduction is seen as a knowledge-extending procedure because theorems are concentrated into basic definitions and a resource-consuming procedure of extraction is needed.

4.4. ATTEMPTS TO REJECT THE SCANDAL

Nevertheless, after his formulation of the distinction between sense and reference,[5] Frege revises the *Grundlagen*'s conception of definitions. In particular, the fruitfulness of good definitions devised in the *Foundations* is substituted by a view that confines their usefulness to an abbreviatory and simplificatory function: after 1884, definitions are, from a logical point of view, "wholly inessential and dispensable (Frege, 1979, p. 208)" and are devoid of any creative power. This radical change in view brings along an equally drastic change in Frege's response to the paradox of analysis. In the Review of Husserl's *Philosophie der Arithmetik I* published in 1894 and in the posthumous text entitled *Logic in Mathematics* dated 1914, Frege paves the way for a psychologistic solution.

4.4.2 Hintikka

A different perspective is held by Hintikka in his *Logic, Language-Games and Information* (1973). In his work, he attacks both the thesis that logic is analytic and the thesis that logic is tautological. As for the first point, Hintikka holds that there exists a class of polyadic truths of first-order logic that are synthetic *a priori*. This claim results from an original definition of the analytic-synthetic distinction, according to which an argument is analytic if and only if it does not introduce new individuals into the discussion, and it is synthetic otherwise. Moreover, Hintikka (1973, p. 137) holds that this definition "approximates rather closely Kant's notion of analyticity" and represents a fair reconstruction in modern terms of Kant's conception.

Although the latter claim might be objected to on several grounds (see Larese, 2020), it cannot be denied that Hintikka's definition is surely Kantian in spirit. To see that this is the case, recall that, according to the definition of the *Critique of Pure Reason*, the concept of the predicate of synthetic judgments is not contained in the concept of the subject. The connection between the two concepts involved, which is necessary for grounding the truth of the judgment, can only be indirect in the sense that it must link the two

---

[5]Between the *Grundlagen der Arithmetik* (1884) and the first volume of the *Grundgesetze der Arithmetik* (1893), Frege introduces his well-known distinction between *Sinn* (sense) and *Bedeutung* (reference, meaning, significance). First presented in the 1891 article *Function and Concept*, the distinction was developed in another paper entitled *Sense and Reference* (1892). Roughly put, the reference of a proper name is the objects it indicates, while its sense is said to be the expression of the name.

concepts to one another by connecting them to a third element. The third element that is always necessary for the truth of synthetic judgments is, for Kant, an object, and the relation between concepts and objects must always be mediated by intuitions, which are representations of individual objects (Kant, 1998, A155/B194 and ff.). Now, the familiarity between Kant's and Hintikka's definitions should be clear: the individuals introduced in synthetic arguments according to Hintikka's definition mirror the intuitions that characterise Kant's synthetic judgments.

Notice that Hintikka's definition is, at the same time, both inspired by Kant in a strong (yet to be measured precisely) sense and applicable, unlike Kant's original definition, to not only categorical judgements but also non-categorical judgements. To be more precise, Hintikka's definition applies to every judgement that can be expressed through the means of modern first-order logic. In taking Kant's original definition, (more or less faithfully) translating it into modern terms and extending it to non-categorical judgments, Hintikka drops one of the hardest charges against the Kantian idea and restores it to its pride of place not only in the history of the concept but also in the contemporary debate. Through Hintikka's insight, a Kantian-inspired definition is also a viable option after the invention of modern first-order logic.

Hintikka's work also provides a formal definition of his analytic-synthetic distinction based on the theory of distributive normal forms for first-order logic and finds in the undecidability of first-order logic the ultimate reason for the thesis that logic is synthetic. Without entering into technicalities, it is sufficient to say that the theory of distributive normal forms for first-order logic is formulated by Hintikka as an extension of the corresponding theory for propositional logic and monadic first-order logic and provides a description of possible worlds, where an upper bound is fixed on the complexity of the configurations of individuals that can be considered together. To this end, each formula F is characterised by the following parameters:

P1. the set of all the predicates occurring in F;
P2. the set of all the free individual symbols occurring in F; and
P3. the maximal length of sequences of nested quantifiers occurring in F, which is called the *depth* of F and is indicated

4.4. ATTEMPTS TO REJECT THE SCANDAL

with d.

Hintikka proves that every first-order formula F with certain parameters can be converted into its distributive normal form with the same parameters or with certain fixed larger ones. As a special case, every constituent with depth d and some parameters P1 and P2 can be converted into a disjunction of constituents, called *subordinate*, with the same parameters P1 and P2 but greater depth $d + e$ for some natural number e.

This observation is crucial for Hintikka's distinction between inconsistent constituents that are *trivially inconsistent* and inconsistent constituents that are *not trivially inconsistent*. While the former are blatantly self-contradictory, the inconsistency of the latter can be detected only by increasing their depth. In other words, it is shown that for every inconsistent constituent of depth d there is some natural number e such that all the subordinate constituents of depth $d + e$ are trivially inconsistent. However, as Hintikka makes clear, due to the Church-Turing result that first-order logic is undecidable, it is not known which depth must be reached for acknowledging that a certain constituent is inconsistent:

> In propositional logic and in monadic first-order logic, distributive normal forms yield a decision method: if a formula has a non-empty normal form, it is satisfiable, and vice versa; it is logically true if and only if its normal form contains all the constituents with the same parameters as it. In view of Church's undecidability result, they cannot do this in the full first-order logic (with or without identity). It is easily seen that this failure is possible only if some of our constituents are in this case inconsistent. In fact, the decision problem of first-order logic is seen to be equivalent to the problem of deciding which constituents are inconsistent. More explicitly, the decision problem for formulae with certain fixed parameters is equivalent to the problem of deciding which constituents with these parameters are inconsistent (Hintikka, 1973, p. 255).

But how can the theory of distributive normal forms provide a distinction between analytic and synthetic inferences? To begin with, Hintikka defines a proof method. To prove G from F, it is necessary to combine the parameters P1 and P2 of F and G, take the maximum, say d, of their depths, and convert F and G into their distributive normal forms F^d and

G^d with the parameters just obtained. Then, F^d and G^d must be expanded by splitting their constituents into disjunctions of deeper and deeper constituents and at each step all the trivially inconsistent constituents must be omitted. Finally, if G follows from F, there will be an e such that all the non-trivially inconsistent members of F^{d+e} obtained by this procedure are among the non-trivially inconsistent members of G^{d+e} obtained through the same procedure.

Now, an inference from F to G is analytic if and only if no increase in depth is needed: the elimination of trivially inconsistent constituents of depth d is sufficient to show that all the constituents of F^d are among those of G^d. On the contrary, an inference from F to G is synthetic if and only if, in order to bring out the desired relationship between F and G, an increase in depth is required: it is necessary to consider configurations of individuals of greater complexity than those that represent the premise and the conclusion of the argument. In other words:

> A sentence F_1 follows from F_0 analytically if and only if the distributive normal form of F_0 will become a part of the normal form of F_1 as soon as trivially inconsistent constituents are eliminated from it (Hintikka, 1973, p. 185).

Moreover, this classification of logical inferences allows Hintikka to give a characterisation of syntheticity as a matter of degree: an inference of G from F is synthetic of degree e if and only if it is necessary to expand the normal forms of F and G by splitting their constituents into disjunctions of depth $d + e$. In other words, the degree of syntheticity of an inference counts the number of individuals that must be included in the initial configurations in order to be able to derive the conclusion from the premise.

Hintikka's work on the syntheticity of first-order logic is deeply connected with his attack against the psychologistic explanation of the highly counterintuitive consequences of the thesis that logic is tautological:

> If no objective, non-psychological increase of information takes place in deduction, all that is involved is merely psychological conditioning, some sort of intellectual psychoanalysis, calculated to bring us to see better and without inhibitions what objectively speaking is already before our eyes. Now most philosophers have not taken to the

4.4. ATTEMPTS TO REJECT THE SCANDAL

> idea that philosophical activity is a species of brainwashing. They are scarcely any more favourably disposed towards the much far-fetched idea that all the multifarious activities of a contemporary logician or mathematician that hinge on deductive inference are as many therapeutic exercises calculated to ease the psychological blocks and mental cramps that initially prevented us from being, in the words of one of these candid positivists, "aware of all that we implicitly asserted" already in the premises of the deductive inference in question (Hintikka, 1973, p. 223).

The dissatisfaction with the solution of the problem raised by the paradox of analysis provided by the logical empiricist movement leads Hintikka to define two notions of information, that he calls "depth" and "surface" information respectively.

"Depth information" is a reconstruction of the notion of semantic information chosen by Bar-Hillel & Carnap (1953) for their theory. An essential feature of depth information is its non-recursive character. This amounts to saying that depth information is not calculable in practice and that in general there is no decision procedure through which the initial distribution of probability can be assigned. Hintikka considers this feature as a major disadvantage of depth information:

> But measures of information which are not effectively calculable are well-nigh absurd. What realistic use can there be for measure of information which are such that we in principle cannot always know (and cannot have a method of finding out) how much information we possess? One of the purposes the concept of information is calculated to serve is surely to enable us to review what we know (we have information about) and what we do not know. Such a review is in principle impossible, however, if our measure of information are non-recursive (Hintikka, 1973, p. 228).

The non-recursive character of depth information is seen by Hintikka as a good reason to react against the conception of information elaborated by Bar-Hillel and Carnap that became the traditional option and to formulate an alternative measure of information that he calls "surface information". The latter is calculable in practice and is more realistic than the depth one, because, unlike depth information, it can be increased by logical deduction: this tool enables us to find that certain alternatives about the world were only apparently viable. But what is surface information about? According to Hintikka (1973, p. 230 and ff.), surface information has a double nature: on the one hand, it is information about reality, because it allows to exclude the existence of (mutually related) individuals; on the other hand, it is conceptual information regarding the conceptual system, that is to say, to the relation between the first-order sentences and the reality they speak of. This observation leads Hintikka to maintain that in first-order logic, due to its undecidability, conceptual information is inseparable from factual information, for the elimination of only apparently consistent constituents, which cannot be effectively isolated, improves our knowledge both of the reality and of our conceptual system.

With this distinction at hand, Hintikka proposes an original solution to the paradox of analysis. While the logical positivists' perspective is vindicated by the recognition that depth information is not increased by logical deduction, the idea that logic is not merely an "intellectual psychoanalysis" is justified by the fact that surface information can be increased during a deduction. The latter kind of information provides an objective and non-psychological sense in which logic is informative.

Hintikka's impressive challenge against the logical empiricist claim that logic is both analytic and tautological cannot be underestimated, but at the same time it is marred by several problems (see Larese, 2023a). The existing literature has mainly focused on one of them: the complexity of the proof procedure. (Rantala & Tselishchev, 1987, p. 89) admit that "as an actual method, the use of normal forms is not very practical", while Lampert, using the results obtained by Nelte (1997, Sect. 4.1), provides a fairly clear picture of the extent of this impracticality. He says:

> Even if one considers only formulas of pure FOL without names and functions, only one binary predicate, and formulas of depth 2, this leads to FOLDNFs with 2512 disjuncts, where each disjunct contains 512 conjuncts. Thus, the length of Hintikka's distributive normal form for even the simple formula $\exists x \exists y Fxy$ is 2^{512}. Merely increasing the depth by one already results in $2^{21+2^{35}}$ possible disjuncts (Lampert, 2017, p. 326).

The extreme difficulty of distributive normal forms for first-order logic is an insurmountable obstacle to using Hintikka's analytic-synthetic distinction in practice.

As Sequoiah-Grayson (2008, p. 88 and ff.) and D'Agostino & Floridi (2009, p. 278) underline, another fundamental problem of Hintikka's proposal is that the class of analytic arguments (or truths) is broader than it might first appear. It includes, beyond many polyadic deductions, the entire set of not only propositional but also monadic arguments. Because propositional and monadic calculi contain only consistent constituents, the inferences included in this set fail to be synthetic and thus increase deductive information. However, is the principle of analyticity and tautologicity of propositional and monadic logic not an "equally disquieting scandal of deduction"? Is Hintikka's thesis not liable to the same accusations that the Finnish logician directed against the Vienna Circle? Is his proposal only a partial solution? Given the probable intractability of Boolean logic, holding that propositional and monadic calculi are analytic and tautological seems to be no less counterintuitive than arguing that full first-order logic is not informative. This observation motivates the approach of Depth-Bounded Boolean Logics.

4.5 The depth-bounded approach

4.5.1 Defusing the scandal

The depth-bounded approach can offer a solution to the debate on the analyticity and tautologicity of logic, by providing a classification of the inferences that are valid in classical propositional logic according to a definition of the analytic-synthetic distinction based on a realistic notion of "holding information".

In Part I we have introduced two infinite hierarchies of approximating logical systems that converge to classical propositional logic. First, consider the logic $\vDash_0$, i.e. the basic layer of either hierarchy. The inferences that are valid in the logic $\vDash_0$ are analytic according to the semantic or meaning-theoretical sense defended by the logical empiricist movement (Sections 4.1 and 4.2): any 0-depth inference is analytical in the sense that whenever the premises are true, the conclusion must also be true by virtue only of the accepted meaning of the logical operators. But while in classical logic the meaning of the logical operators is defined through the standard truth-conditions, which hinge on the classical information-transcending notions of truth and falsity as primary semantic notions, as we have seen in Chapter 1, in Depth-Bounded Boolean Logics the meaning of the logical operators is specified instead through the notions of "holding the information" that a given sentence is true, respectively false, where these notions are construed in a practical, operational sense.

In Section 4.2, we have shown that the principle of semantic analyticity, together with the paradox of analysis, leads to the principle of informational analyticity or tautologicity. The logic $\vDash_0$ makes no exception and, as a first approximation, we have that any 0-depth inference is tautological in the sense that whenever we possess the information that the premises are true, we thereby possess the information that the conclusion is true. But here, as a result of the shift from the classical and information-transcendent primary notions to the their epistemic counterparts, the notion of "possessing the information" means "*actually* possessing the information". Thus, we can say that the logic $\vDash_0$ satisfies also the following *strict* informational sense of analyticity or *strict* tautologicity:

> *Strict informational analyticity or strict tautologicity.* An inference is analytical in the strict informational sense, or strictly tautological if, and only if, whenever we *actually* hold the information that the premises are true, we thereby *actually* hold the information that the conclusion is true.

This strict informational sense of "analytic" helps clarifying the Cohen-Nagel paradox (see Section 4.2), which can be construed as the following argument:

1. Classically valid inferences are analytic in the semantic sense

2. If an inference is analytic in the semantic sense, then it is informationally trivial

∴ Classically valid inferences are informationally trivial

where the conclusion clashes with the fact that, owing to the lack of a feasible decision procedure, deductive inference does reduce our practical uncertainty even in the domain of propositional logic. This is a genuine conflict only if "informationally trivial" is construed as "analytic in the strict informational sense": any agent who holds the information that the premises are

4.5. THE DEPTH-BOUNDED APPROACH

true, thereby holds the information that the conclusion is true, where "holding the information" is intended in the practical, operational sense. Then, since classical propositional logic is NP-hard, it is highly implausible that a feasible decision procedure will ever be found. Hence, there is no guarantee that, for any agent a who grasps the classical meaning of the logical operators, a holds the information that A is true, whenever a holds the information that the sentences in Δ are true and A is a classical consequence of Δ.

Where is the catch? The first premise of the "paradox" is usually shown to be true by defining the meaning of the logical operators via the standard truth-conditions, which hinge on the classical information-transcending notions of truth and falsity as primary semantic notions. However, there is a patent *mismatch* between these notions and the central epistemic notions underlying the strict informational sense of "analytic", namely *holding (actually possessing) the information* that a sentence is true, respectively false, which is the only sense that makes the argument sound paradoxical. Hence, if the first premise is recognized as true on the grounds of the classical semantics for the logical operators, the second premise is far from being compelling, or even plausible, and so the whole argument is not much of a paradox, after all.

To sum up, we have seen that the depth-bounded approach provides a definition of the analytic-synthetic distinction that can be interpreted as a restriction on the properties of analyticity and tautologicity as understood by the logical empiricist movement. On the one hand, semantic analyticity is based on the informational, rather than the classical, meaning of the operators; on the other hand, informational analyticity involves information that is possessed not only in principle, as in the classical version of the principle, but also in practice. A result of this restriction is that while for logical empiricism the principles of analyticity and tautologicity of logic apply to classical logic as a whole, according to the depth-bounded approach these properties apply only to a subset of the inferences that are classically valid, i.e., inferences valid in the logic $\vDash_0$. We might say that the logic $\vDash_0$ vindicates the fundamental theses held by the logical empiricist movement on the epistemological status of logic, but only in a restricted domain.

At the same time, the inferences that are classically valid, but not valid in the logic $\vDash_0$ appear to vindicate the opposite view, namely, that classical logic is synthetic *a priori*, which has been famously attacked by the logical empiricist movement. The notion of synthetic *a priori* provided by the depth-bounded approach is based on two observations. First, the informational meaning of the logical operators is not sufficient to obtain all the inferences that are classically valid; on the contrary, what is needed is a certain number of nested pieces of virtual information. This means that the approximations $\vDash_k$, for $k \geq 0$, validate also inferences that are not analytic in the semantic sense as specified by the informational meaning of the logical operators. Second, the use of virtual information, which is necessary to retrieve all the classically valid inferences, goes beyond the information that is actually possessed. Thus, the logics $\vDash_k$, for $k \geq 0$, validate also inferences that are not analytic in the strict informational sense, i.e. the informational sense obtained by specifying the meaning of "possessing the information" in compliance with a manifestability requirement (see Chapter 1). Moreover, following Hintikka's insight (see Section 4.4), the depth-bounded approach as a whole makes the analytic-synthetic distinction a *matter of degree*. This can be seen as a further articulation of Hintikka's plea for a more realistic view that focuses on "the information we actually possess (as distinguished from the information we in some sense have potentially available to us) and with which we can in fact operate" (Hintikka, 1973, p. 229), namely what he calls "surface information" as opposed to "depth information". Hintikka introduced this distinction to defend the informativeness of classical first-order logic, which he related to its undecidability. While in his view classical propositional logic is really tautologous, we claim that a similar problem is raised by its probable intractability and that an even more realistic view can be construed in terms of the depth-bounded approach. The greater the number of nested pieces of virtual information used in a certain inference, the greater the depth of that inference. However, against Hintikka's proposal, the depth-bounded approach classifies as synthetic at an increasing degree all classical propositional inferences that are regarded as "hard" by real world agents (human, artificial or hybrid) depending on the available resources. In so doing, it vindicates not only the layman's intuition, but it also takes into serious account (and possibly sheds light on) the widely assumed intractability of Boolean logic.

How does the notion of synthetic *a priori* provided by the depth-bounded approach relate to Kant's foundational characterisation of this notion (see

4.5. THE DEPTH-BOUNDED APPROACH

Section 4.1)? The use of virtual information, that is not contained in the data and so may not be actually held by any agent who holds the information carried by the data, appears to qualify the arguments in which it is used as "synthetic" in a sense close to Kant's, in that it forces the agent to consider potential information that is not included in the information "given" to him:

> Analytic judgments (affirmative ones) are thus those in which the connection of the predicate is thought through identity, but those in which this connection is thought without identity are to be called synthetic judgments. One could also call the former **judgments of clarification**, and the latter **judgments of amplification**, since through the predicate the former do not add anything to the concept of the subject, but only break it up by means of analysis into its component concepts, which were already thought in it (though confusedly); while the latter, on the contrary, add to the concept of the subject a predicate that was not thought in it at all, and could not have been extracted from it through any analysis (Kant, 1998, A 6–7/B 10–11). [...]
> In the analytic judgment I remain with the given concept in order to discern something about it [...] In synthetic judgments, however, I am to go beyond the given concept in order to consider something entirely different from what is thought in it [...] (Kant, 1998, A 154/B 193).

It is the necessity of considering something that goes *beyond what is given to us* in the premises, i.e. beyond our actual information, that makes such inference steps non-analytic in a sense analogous to Hintikka's (and Kant's).[6]

By contrast, analytical inferences are those which are recognized as sound via steps which are all "explicative", that is, descending immediately from the meaning of the logical operators, as given by the necessary and sufficient conditions expressed by the elimination and introduction rules, while synthetic ones are those that are "augmentative", involving some intuition that goes beyond this meaning, i.e., involving the consideration of virtual information.

In a recent work, Mendonça (2023, p. 23) holds that the appeal to Kantianism is misleading: "the informativeness of demonstrations involving dischargeable hypotheses" should be explained "without appealing to Kantian-

---
[6] See also (D'Agostino & Floridi, 2009), (Larese, 2022) and (Larese, 2023b) on this point.

ism", the reason being that "the informativeness of such proofs does not relate to the construction of conceptual intuitions". This claim prompts for two observations. First of all, the analytic-synthetic distinction provided by Depth-Bounded Boolean Logics has a strong Kantian flavour, although it is not intended to be a philologically accurate translation in modern terms of Kant's position. In particular, virtual information plays the role that intuitions have in Kant's conception: both of them go beyond the concept of the subject (or beyond what is contained in the premises), cannot be found through analysis and are the essential elements of a preparatory synthetic phase that is followed by an analytic proof. But it is clear that, while the *role* of virtual information is surely Kantian, its *nature*, as (Mendonça, 2023) seems to suggest, cannot be compared to the notion of intuitions as conceived in the *Critique*.

Second, this appeal to Kantianism is not meant to be an *explanation* of the informativeness of k-depth Boolean Logics. On the contrary, proofs making essential use of virtual information are informative because their conclusion does not follow from the informational meaning of the logical connectives and, as a consequence, an agent is required to visit epistemic states that are richer than the one they possess when they are informed fo the premises. Therefore, rejecting the claim that the notion of syntheticity as the use of virtual information is Kantian in spirit does not affect the claim that k-depth Boolean Logics are informative. For the same reason, (Mendonça, 2023)'s interesting suggestion that virtual information can be interpreted as information introduced by the audience of a demonstration to cooperate with the deductive process cannot be taken as a *justification* of the informativeness of such proofs.[7]

4.5.2 Semantic information revisited

Let us now conclude by outlining a weaker notion of semantic information according to which the basic layer of our approximations is truly informationally trivial.

The set S of all shallow information states discussed in Chapter 1 (in-

---

[7]Moreover, the claim that "dialogical logics can furnish a more appropriate framework for logical investigation that the traditional monological alternatives" is simply "assume[d]" by (Mendonça, 2023, p. 9).

4.5. THE DEPTH-BOUNDED APPROACH

tended as a set of pieces of information represented by pairs of the form $\langle A, i \rangle$ where i is 1 or 0) is naturally ordered by set inclusion. Every non-empty subset $\mathcal{P}$ of $\mathcal{S}$ has a meet in $\mathcal{S}$ given by $\bigcap \mathcal{P}$. On the other hand, two information states might not have a join in $\mathcal{S}$. Indeed, if two information states are mutually inconsistent they have no upper bounds in $\mathcal{S}$. Observe also that, even when two information states have a join in $\mathcal{S}$, this is not, in general, their set union. For example, the join in $\mathcal{S}$ of two information states containing, respectively, $\langle p \vee q, 1 \rangle$ and $\langle p, 0 \rangle$, must contain also the signed sentence $\langle q, 1 \rangle$ that may not be contained in either of them. Given a subset $\mathcal{P}$ of $\mathcal{S}$, let $\mathcal{P}^u$ be the set of all upper bounds of $\mathcal{P}$ in $\mathcal{S}$. Then, $\mathcal{P}$ has a join in $\mathcal{S}$ whenever $\mathcal{P}^u$ is non-empty, and this is given by $\bigcap \mathcal{P}^u$. Now, since $\mathcal{S}$ itself has no upper bounds in $\mathcal{S}$, this ordering is topless. Let $\top$ be the set of *all* pieces of information (which is *not* an information state) and let $\mathcal{S}^* = \mathcal{S} \cup \{\top\}$. Then $(\mathcal{S}^*, \subseteq)$ is a complete lattice, where the meet of an arbitrary subset $\mathcal{P}$ of $\mathcal{S}^*$ is given by $\bigcap \mathcal{P}$, while its join is equal either to the top element $\top$, if $\mathcal{P}^u$ is empty, or to $\bigcap \mathcal{P}^u$ otherwise.

Now, the *shallow information* carried by a *sentence* A, $\mathrm{INF}(A)$ can be defined as

(4.2) $$\mathrm{INF}(A) = \bigcap \{S \in \mathcal{S} \mid T\,A \in S\}.$$

More generally, the *shallow information* carried by a *set* Γ of sentences can be defined as

(4.3) $$\mathrm{INF}(\Gamma) = \bigcap \{Y \in \mathcal{S} \mid \Gamma \subseteq Y\}.$$

Observe that, since $\bigcap \emptyset = \top$, (4.3) yields $\mathrm{INF}(\Gamma) = \top$ whenever Γ is 0-depth inconsistent, for there is no $Y \in \mathcal{S}$ that may include Γ. Recall that $\top$ is not an information state, but only denotes a situation in which all information is "suspended" and can be rather interpreted as a call for revision. So $\top$ is conceptually distinct from the *empty information state*, that is, the partial valuation that is undefined for all formulae. However, an agent whose informational situation is described by $\top$ holds no genuine information just as any agent whose information state is empty. Then, in order to be informative for an agent a, a (set of) sentence(s) must be 0-depth consistent.

This requirement of 0-depth consistency (not classical consistency) can be seen as a substantial mitigation of the "veridicality thesis" put forward by Luciano Floridi as a solution to the BHC paradox. Even if one is not willing to endorse the somewhat controversial view that "information encapsulates truth", (Floridi, 2004) and (Floridi, 2011, Chapters 4–5), one can still maintain that a minimal interpretation of "holding information" is one that satisfies the requirement that no agent may hold information that is explicitly inconsistent. And if a set of sentences Γ is 0-depth inconsistent, no agent a can "hold the information" that all the sentences in Γ are true, because adding Γ to a's current information state would destroy the latter as an information state.

The *informativeness of Γ for an agent a*, $\iota_a(\Gamma)$ can be characterized as follows:

(4.4) $$\iota_a(\Gamma) = \text{INF}(\text{INF}(S_a \cup \Gamma) \sim S_a),$$

where S_a is the current information state of a. Again, it follows from (4.4) that $\iota_a(\Gamma) = \top$ whenever Γ is 0-depth inconsistent.

On the basis of the above definitions, the 0-depth consequence relation can be equivalently defined as follows:

$\Gamma \vdash \varphi$ if and only if $\text{INF}(\varphi) \subseteq \text{INF}(\Gamma)$.

Hence, $\vdash$ is informationally trivial, in that every agent that actually holds the information that the premises are true must thereby hold the information that the conclusion is true, or equivalently, the shallow semantic information carried by the conclusion is included in the shallow semantic information carried by the premises. The latter wording covers the limiting case in which the shallow information carried by the premises is $\top$ which do not qualify as genuine information ($\top$ is not an information state).

It may be objected that the consequence relation $\models$ is still "explosive" when Γ is 0-depth inconsistent, for there is no information state for a that contains $\langle A, 1 \rangle$ for all $A \in \Gamma$. So, if Γ is 0-depth inconsistent $\Gamma \models A$ for every sentence A. Similarly, if we follow the informational definition of 0-depth consequence just given, when Γ is 0-depth inconsistent, $\text{INF}(\Gamma) = \top$ and so, for every A, $\text{INF}(A) \subseteq \text{INF}(\Gamma)$. However, the problem raised by this kind of explosivity is far less serious than the similar problem for the classical consequence relation. For, we can detect that the premises are 0-depth inconsistent in feasible time. Unlike hidden classical inconsistencies, that may be hard to discover even for agents equipped with powerful

4.5. THE DEPTH-BOUNDED APPROACH

(but still bounded) computational resources, 0-depth inconsistency lies, as it were, on the surface. So, we always have feasible means to ensure that our premises are 0-depth consistent, in which case the consequence relation $\models_0$ is *not* explosive, even if these premises are *classically inconsistent*.

We stress again that our definition of information state and shallow information do not require that information "encapsulates truth", nor do they even require that it "encapsulates consistency", but only that information "encapsulates shallow (0-*depth*) consistency". According to this characterization, $\models$ is informationally trivial by definition, and this is in accordance with the tenet that analytic inferences are utterly uninformative. The valid inferences of $\models$ are only a subclass of the classically valid inferences and their validity can be recognized in feasible time. These are the "easy" inferences that (nearly) everybody learns to make correctly in the very process of learning the meaning of the logical operators.

Conclusion.
The depth-bounded paradigm

The approach presented in this book provides a solution to the approximation problem in classical propositional logic which, at the same time, defuses the scandal of deduction and the anomalies of the received view on the relation between logic and information. An inference is uninformative or "analytic", if the conclusion follows from the premises by virtue of the sole informational meaning of the logical operators; otherwise, it is informative, or "synthetic" at various degrees. Moreover, the graded notion of logical depth it provides has turned out to be useful in some AI application areas, such as formal argumentation, and is a promising asset in all contexts in which the idea of bounded rationality applies.

We start by briefly discussing ideas and methods that are most closely related to the semantic and proof-theoretical framework presented here.

KE and KI. Let us consider a propositional language $\mathcal{L}$ consisting only of the four standard Boolean operators $\wedge, \vee, \neg, \rightarrow$. We may restrict our attention to the *intelim*-sequences that contain only applications of the elimination rules of Tables 1.5 and 1.6, that we may call "E-sequences". An *eliminative refutation* of Γ is a closed elimination sequence for $\{T\,B \mid B \in \Gamma\}$.

On the other hand, we may also restrict our attention to the *intelim*-sequences that contain only applications of the introduction rules of Γ, that we may call "I-sequences". An *introductory proof* of A from Γ is an introductory sequence for $\{T\,B \mid B \in \Gamma\}$ that contains $T\,A$.

Each of these two restricted notion is turned into a complete system for classical propositional logic by adding a single structural rule, corresponding to the classical *Principle of Bivalence*, that allows us to split a sequence into two branches that are developed in parallel:

(RB) $$\frac{}{T A \mid F A}$$

A *KE-tree* for a set Γ of a signed formulas is a tree whose nodes result from applications of the *elimination rules* of Tables 1.5 and 1.6, starting from the signed formulas in Γ. A *KE-refutation* of a set Γ of formulas is a closed KE-tree for $\{T B \mid B \in \Gamma\}$, namely one in which all branches are *closed*, i.e., contain a pair of conjugate signed formulas. KE is a complete refutation system for classical propositional logic and remains complete if the applications of the branching rule RB are restricted to subformulae of the formulas in the initial set Γ.

A *KI-tree* for a set Γ of a signed formulas is a tree whose nodes results from applications of the *introduction rules* of Tables 1.5 and 1.6, starting from the signed formulas in Γ. A *KI-proof* of a A from Γ is a KI-tree for $\{T B \mid B \in \Gamma\}$ such that $T A$ occurs in all *open* branches. KI is a complete proof system for classical propositional logic and remains complete if the applications of the branching rule RB are restricted to atomic formulas occurring either in the assumptions or in the conclusion. Moreover, the applications of the introduction rules can be restricted, without loss of completeness, to subformulae of the assumptions or of the conclusion.

The system KE and KI were introduced by Marco Mondadori (Mondadori, 1988a,b,d) and their relative complexity was later investigated in D'Agostino (1990, 1992); D'Agostino & Mondadori (1994); Mondadori (1995); D'Agostino (1999), showing that both methods dominate Smullyan's analytic tableaux in terms of the p-simulation relation (Cook & Reckhow, 1974).

It has been shown (Mondadori, 1989a; D'Agostino, 2005; D'Agostino *et al.*, 2006) that combining KE and KI leads to a sort of "natural deduction" system in which the inferential behaviour of the logical operators is regulated by introduction and elimination rules that mirror their classical meaning — while Gentzen's original rules are better seen as an explication of their intuitionistic meaning — and satisfy Prawitz's inversion principle. These operational rules require no manipulation of virtual information (no "discharge" of provisional assumptions), the latter being governed only by

4.5. THE DEPTH-BOUNDED APPROACH

the structural rule RB. Normalization of proofs and subformula property for this system of natural deduction are shown in D'Agostino (2005). Given the presence of the introduction rules, this natural deduction system can be used, indifferently, as a refutation method or as a proof method for classical propositional logic.

In all these tree methods, the operational rules are not branching and the only branching rule is the structural rule RB. It is therefore quite natural to investigate the subsystems that result from limiting the applications of this single structural rule. A suggestion in this sense for the KE system can be found in D'Agostino & Mondadori (1994, Prop. 5.6); see also D'Agostino (1999, Prop. 46). KE with restricted bivalence has been more thoroughly investigated in Finger (2004a), Finger (2004b) and Finger & Gabbay (2006). A similar idea for the natural deduction system that results from combining KE and KI is discussed in D'Agostino (2005). In general, the depth-bounded consequence relations that can be characterized simply by limiting the nested applications of the RB-rule in the combined KE/KI system are the *weak k-depth* consequence relations discussed in Section 2.1. The relationship between KE, KI and Gentzen-style natural deduction has been recently discussed at length in Indrzejczak (2010) with special attention to the complexity of proof-search methods.

Stålmarck's method. Stålmarck's method is a patented algorithm for tautology checking that was first described in Stålmarck (1992). The method has proved to be quite useful for several industrial and civil applications involving the efficient verification of critical properties that are typically expressed by means of (complex) Boolean formulas. The main heuristic principles underlying this algorithm are very similar to those underlying the KE/KI approach: perform Boolean inference by means of non-branching operational rules augmented by a single branching rule (closely related to the RB rule of KE and KI), and delay the application of the latter as long as possible. The two research programmes developed independently and in parallel, addressing different communities. Perhaps their connection was partially obscured by the fact that, in its original presentation, Stålmarck's algorithm applied to formulas in implication normal form and was based on rewrite rules whose format was apparently very distant from that of typical inference rules in more conventional proof-theoretical frameworks, such as analytic tableaux or natural deduction. The gap has been bridged in (Sheeran & Stålmarck, 2000; Björk, 2003), where the connection with KE and KI, as well as the contrast with Gentzen style calculi, are clearly and explicitly stated. (See also Hähnle (2001, §5.4) on the connection between KE/KI and Stålmarck's method).

In essence, the proof-theoretical backbone of Stålmarck's method can be described as a natural deduction system — like the one described in D'Agostino (2005) and D'Agostino & Floridi (2009)[8] and generalized in Section 2.2.2 of this book — combining the elimination rules of KE, the introduction rules of KI and a variant of RB called "dilemma" — see Sheeran & Stålmarck (2000, Section 3.7); Björk (2003, pp. 17–20). Like in KE/KI the latter is the only branching rule and the only one that involves the introduction and subsequent discharge of (complementary) virtual assumptions. Stålmarck's method is a refutation algorithm exploiting the fact that the unbounded system — where the number of nested applications of RB is not limited — enjoys the subformula property (see D'Agostino (2005) for a procedure to turn arbitrary proofs into ones that enjoy the subformula property), so that the formulas occurring in the conclusion of the introduction rules and in the RB rule can be restricted to (weak) subformulae of the initial assumptions or of the conclusion. However, the algorithm adds to this proof-theoretical hardcore two kinds of enhancements: *branch merging*, that allows to unify adjacent branches by deleting all the formulas that do not occur in both of them, and *formula relations*, recording the fact that two formulas must have the same (unknown) truth value.

If we leave formula relations aside, Stålmarck's algorithm for 0-depth saturation is a procedure for recognizing the *intelim*-inconsistency of formulas in implication normal form. It follows from the subformula property that allowing for unrestricted applications of the introduction rules does not generate any new valid inference and so, given completeness, the algorithm is sound and complete for 0-depth inconsistency, defined in terms of the Tarskian consequence relation $\models_0$, which is characterized by the semantic approaches of Chapter 1. Stålmarck's k-depth algorithm corresponds to k-depth inconsistency defined in terms of the relations $^+\models_k^{sub,sub}$ and $\vdash_k^{sub,sub}$ of Section 2.2.2. For $k > 0$, $\vdash_k^{sub,sub}$ is not equivalent to $\vdash_k^{F,sub}$ and so, un-

---

[8] In these papers the authors were not aware of the exposition in Sheeran & Stålmarck (2000).

4.5. THE DEPTH-BOUNDED APPROACH

like the 0-depth case, restricting the conclusions of the introduction rules to subformulae of the initial formulas involves some loss of inferential power. (Recall also that for any $k > 0$, $\vdash_k^{F,F}$ is full Boolean Logic, see Proposition 2.1.8.)

With the addition of formula relations the situation is less transparent. Although these enhancements may significantly improve the efficiency of the k-depth saturation procedure, for each fixed k, they seem not to fit the "harmonic" pattern of introductions and eliminations that is exhibited by the "pure" rules, which is likely to affect the very possibility of a well-behaved theory of meaning for the logical operators. So, the semantic characterizations given in this paper do not appear to be easily adaptable to provide similar characterizations for Stålmarck's k-depth saturation. From a theoretical viewpoint, it seems more convenient to investigate the systems based on the pure rules and treat other enhancements as shortcuts for refutations that are more aptly presented in a higher-depth system. In any case, we believe that the proof-theoretical and semantic framework presented here can be useful also to further investigate the properties of Stålmarck's procedures.

The approach of Depth-Bounded Boolean Logics has given rise to an articulated research programme at the crossroad of philosophy, mathematical logic and computer science. The main ideas explored in the present book have been applied and extended in recent works and are at the core of ongoing and future research projects. This section provides the status of the art, as well as some bibliographical reference.

First-order logic. A natural extension of Depth-Bounded Boolean Logics is towards the theory of quantification. Larese (2019, Chapter 5) provides some exploratory considerations in this direction, while D'Agostino *et al.* (2021) lay the foundations for extending the informational semantics for the Boolean operators to the standard quantifiers and give a preliminary definition of a depth-bounded natural deduction system for classical first-order logic. To this end, the authors employ Hintikka's notion of syntheticity as the introduction of new individuals into the discussion (see Section 4.4) and offer an alternative formalisation thereof. A comparison between Hintikka's original formalisation through the theory of distributive normal form and the resulting unified treatment of classical first-order logic is carried out in Larese (2023a). Central topics of present and future research in the field of depth-bounded quantification include proving the normalisation theorem and results in the fields of computability and computational complexity.

Modal logic. Pardo (2022) concentrates on the informational approach to classical logic to study a modal extension for reasoning about blueprints for propositional proofs. The motivation of his work lies in distributed reasoning applications, such as query-answering and consistency checks in a shared or distributed data base. In the resulting hierarchy of tractable logics, called UBBL, proof complexity is measured by the number of RB-formulas, understood as a sort of blueprint of proofs, rather than their depth.

Epistemic logic. Larese (2019, Chapter 6) focuses on the problem of logical omniscience and, using the well-known muddy children puzzle as a case study, provides the semantic bases to defining hierarchies of approximations to modal epistemic logics similar to Depth-Bounded First-Order Logics. In this framework, every agent is characterised by two parameters, her set of initial information and her depth, and a distinction is made between three scenarios: i) each agent ignores the other agents' depth; ii) each agent knows the other agents' depth as the result of a private announcement; iii) each agent knows the other agents' depth as the result of a public announcement (common knowledge). These preliminary investigations have been developed further. A forthcoming paper simplifies the Kripke-style semantics of the previous work and defines a proof-system based on the tableaux rules for Depth-Bounded Boolean Logics.

Multi-agents systems. Multi-agents settings have been explored in Cignarale & Primiero (2021) and Larotonda & Primiero (2023). The former starts from the observation that Depth-Bounded Boolean Logics model only the case in which virtual information is simulated by an agent's inner inferential abilities and that in many practical applications, such information is communicated. Consequently, the authors propose MA-DBBL, a multi-agent version of Depth-Bounded Boolean Logics, extending the latter with an operator of information sharing between agents, and one expressing information held by every agent in the system. In the latter paper, the authors stress that MA-DBBL can only model highly collaborative settings, where every

4.5. THE DEPTH-BOUNDED APPROACH

agent transmits every information she possesses. Accordingly, they present DBBL-BI$_n$, a variant logic equipped with a relational semantics that accounts for a multi-agent system where agents are ordered hierarchically and have access to increasingly extensive information states. The underlying idea is that these hierarchies are defined among agents with shared competencies but with different access to relevant information. Moreover, an application in the context of answer set programming is provided in Baldi *et al.* (2021). This work presents a limited-depth version of the popular ASP system *clingo*, tentatively dubbed *k-lingo* after the bound k on virtual information.

Non-classical propositional logics. In his thesis, Solares-Rojas (2022) shows that the Depth-Bounded Boolean approach can be naturally extended to useful non-classical logics such as first-degree entailment (FDE), the logic of paradox (LP), strong Kleene logic (K$_3$) and intuitionistic logic (IL). To this end, he introduces a KE/KI-style system for each of those logics, which naturally yields a hierarchy of tractable depth-bounded approximations to the respective logic. Moreover, he shows that each hierarchy approximating FDE, LP and K$_3$ admits of a 5-valued non-deterministic semantics, whereas, paving the way for a semantical characterization of the hierarchy approximating IL, he provides a 3-valued non-deterministic semantics for the full logic. D'Agostino & Solares-Rojas (2022) and D'Agostino & Solares-Rojas (2024) offer a detailed exposition of the above-mentioned motivations and formal systems, while adding important results such as subformula property, tractability, soundness and completeness.

Probability. A series of papers investigates the application of the Depth-Bounded Boolean approach to the theory of probability. Baldi *et al.* (2020), expanding on the ideas first introduced in Baldi & Hosni (2020), define depth-bounded belief functions, a logic-based representation of quantified uncertainty, which give rise to a hierarchy of increasingly tighter lower and upper bounds over classical measures of uncertainty. This paper also provides the conditions under which Dempster-Shafer Belief functions and probability functions can be represented as a limit of a suitable sequence of Depth-bounded Belief functions. Baldi & Hosni (2021) introduce sequences of qualitative belief structures and identify the conditions under which i) a qualitative sequence approximates a qualitative probability; and ii) a qualitative probability can be approximated. Baldi & Hosni (2023) supersede the previous paper by putting the hierarchy of depth-bounded belief functions in correspondence with an increasingly higher ability to resolve uncertainty beyond the information actually held by the agent and leading in the limit to additive measures. Furthermore, they show that under rather palatable restrictions, these approximations of probability lead to uncertain reasoning which, under the usual assumptions in the field, qualifies as tractable.

Appendix A

Proofs and algorithms

A.1 Proof of Proposition 1.4.3

We now sketch a proof Proposition 1.4.3. We can identify a partial valuation v with the set S_v of signed formulae defined as follows:

$$S_v = \{T\,A \mid v(A) = 1\} \cup \{F\,A \mid v(A) = 0\}.$$

Conversely, given a set S of signed formulae such that for no formula A, $T\,A$ and $F\,A$ are both in S, the partial valuation v_S is defined as follows:

$$v_S(A) = \begin{cases} 1 & \text{if } T\,A \in X \\ 0 & \text{if } F\,A \in X \\ \bot & \text{otherwise.} \end{cases}$$

Say that a set of S-formulae is *intelim saturated* if it is closed under the intelim rules and does not contain any pair of conjugate S-formulae of the form $T\,A$ and $F\,A$.

Observe that, according to our definitions in Section 1.3 a valuation module α is stable, namely closed under SCP, if and only if α is closed under all the intelim rules with premises in S_α. Moreover, a valuation is a shallow information state if and only if all its modules are stable. It then follows that:

Lemma A.1.1. *(1) If a partial valuation v is a shallow information state, then S_v is intelim saturated. (2) If S is intelim saturated, then v_S is a shallow information state.*

Let us now turn to Proposition 1.4.3. As for the "if" direction, suppose $X \vdash_0 \varphi$ and let v be any shallow information state that satisfies X. Then, by Lemma A.1.1, S_v is an intelim saturated set that includes X and so contains φ. Therefore, v satisfies φ. If $X \nvdash_0$, i.e., X is 0-depth refutable, no intelim saturated set includes X and therefore, by the same lemma, no shallow information state satisfies X, that is, $X \models_0$. For the "only if" direction, if $X \nvdash_0 \varphi$, X is 0-depth consistent and the set $S = \{\psi \mid X \vdash_0 \psi\}$ is intelim saturated that includes X and does not include φ. Then, by the above lemma again, v_S is a shallow information state that satisfies X, but does not satisfy φ. Hence, $X \nvDash_0 \varphi$.

A.2 Proof of Proposition 1.4.15

Proof. Let π be a strongly non-redundant intelim proof of φ from X (refutation of X). Suppose there are S-formulae that are not signed subformulae of the premises or of the conclusion and let us call such formulae *spurious*. Let ψ be a spurious S-formula of maximal logical complexity. Clearly, $\psi \notin X \cup \{\varphi\}$. Then, ψ must result from the application of an inference rule. It cannot result from the application of an elimination rule, otherwise π would contain a more complex spurious formula, namely the major premise of this elimination. Therefore ψ must be the conclusion of an introduction. Since π contains no idle occurrences of formulae, either (i) ψ is used as a premise of a rule application, or (ii) it is used to close the sequence, and so ψ and $\overline{\psi}$ both occur in π. As for (i), observe that ψ cannot be a premise of an introduction or the minor premise of an elimination, otherwise there would be again a more complex spurious formula in $\mathcal{T}$. Moreover, ψ cannot be used in π as major premise of an elimination, otherwise ψ would be a detour and, by Lemma 1.4.11, π would be redundant, against the hypothesis. As for (ii), observe that neither ψ nor $\overline{\psi}$ can be conclusions of eliminations, otherwise π would contain a spurious S-formula of greater complexity, namely the major premise of the elimination. Finally, they cannot be both conclusions of introductions, since in this case π would be redundant by Fact 1.4.12. □

A.3 Tractability

Corollary 1.4.17 guarantees the existence of a decision procedure for the intelim logic by delimiting the space of the signed formulae that have to be taken into consideration as possible conclusions of the introduction rules. In this section we show that the decision problem for the intelim logic is tractable.

Definition A.3.1. *The* subformula graph *for Γ is the oriented graph $\langle V, E \rangle$ such that $V = \text{sub}(\Gamma)$ and $\langle A, B \rangle \in E$ if and only if A is an immediate subformula of B.*

Definition A.3.2. *A G-*module *is any subgraph M of G whose set of nodes is an $\mathcal{L}$-module, i.e., consists of a formula with all its immediate subformulae. The* top formula *of M is the top formula of the underlying $\mathcal{L}$-module.*

Definition A.3.3. *A* labelled subformula graph *for Γ is a pair $\langle G, \lambda \rangle$, where G is the subformula graph for Γ and λ is a partial function, called the* labelling function, *from the vertices of G into $\{0, 1\}$.*

Definition A.3.4. *An* intelim graph *for Γ based on Δ, with $\Delta \subseteq \Gamma$ is a labelled subformula graph $\langle G, \lambda \rangle$ for Γ satisfying the following conditions:*

1. *for every $A \in \Delta$, $\lambda(A) = 1$;*

2. *for every A in $\text{sub}(\Gamma)$ such that $A \notin \Delta$, $\lambda(A)$ is defined and equal to 1 (or 0) only if there are $B_1, \ldots, B_k$ in $\text{sub}(\Gamma)$ such that $T\,A$ (or $F\,A$) follows from*

$$\{T\,B_i \mid 0 \leq i \leq k, \lambda(B_i) = 1\} \cup \{F\,B_i \mid 0 \leq i \leq k, \lambda(B_i) = 0\}$$

by an application of an intelim rule.

The initial *intelim graph for Γ based on Δ is the intelim graph for Γ based on Δ such that $\lambda(A) = \bot$ for all $A \notin \Delta$. An intelim graph for Γ is* completed *if it satisfies also the converse of (2).*

Simple decision procedures for intelim refutability and deducibility are illustrated in Algorithms A.3.1–A.3.2, and consist in building the initial intelim graph for the set of formulae that are mentioned in the specification of the problem (just the assumptions for refutation problems, the assumptions plus the conclusion for deduction problems) and turning it into a complete intelim graph by saturating it in accordance with the intelim rules. Both algorithms call the subroutine Expand described in Algorithm A.3.3. The latter, in turn, calls the subroutine Apply_Intelim described in Algorithm A.3.4.

The symbol "$\top$" stands for the "inconsistent labelling function". This is only a way of speaking to mean that there is no labelling function consistent with the intelim rules.

The correctness of both decision procedures follows from the fact that they return true if and only if the set $\{T\,A \mid \lambda(A) = 1\} \cup \{F\,A \mid \lambda(A) = 1\}$ is intelim saturated. We omit a detailed proof and just briefly discuss their complexity. For each $x \in \{0, 1\}$, let $\text{sign}(x) = T$ iff $x = 1$ and

---

Algorithm A.3.1 Decision Procedure for intelim refutability

Require: A finite set Γ of formulae;
1: build the subformula graph G for Γ;
2: set $\lambda(A) = 1$ for each $A \in \Gamma$;
3: set $\lambda(B) = \bot$ for each B such that $B \notin \Gamma$;
4: set expanded_graph = Expand($\langle G, \lambda \rangle$);
5: **if** expanded_graph = $\langle G, \top \rangle$, **then**
6:     **return true**;
7: **else**
8:     **return false**.
9: **end if**

$\text{sign}(x) = F$ iff $x = 0$. Recall also that, for $S \in \{T, F\}$, $\text{val}(S) = 1$ iff $S = T$ and $\text{val}(S) = 0$ iff $S = F$. Let $|A|$ denote the *size* of the formula A, i.e., the total number of occurrences of symbols in A. The size of a finite set Γ of formulae is defined as $\sum_{A \in \Gamma} |A|$ and denoted by $|\Gamma|$.

The input for Algorithm A.3.1 is a list of all the formulae in Γ, while the input for Algorithm A.3.2 is a pair consisting of a list of all the formulae in Γ and the formula A. Let n be the total size of the input, namely $O(|\Gamma|)$ for Algorithm A.3.1 and $O(|\Gamma \cup \{A\}|)$ for Algorithm A.3.2.

A.3. TRACTABILITY

Algorithm A.3.2 Decision Procedure for intelim deducibility

Require: A finite set Γ of formulae and a formula A;
1: build the subformula graph G for $\Gamma \cup \{A\}$;
2: set $\lambda(A) = 1$ for each $A \in \Gamma$;
3: set $\lambda(B) = \bot$ for each B such that $B \notin \Gamma$;
4: set expanded_graph = Expand($\langle G, \lambda \rangle$);
5: **if** expanded_graph = $\langle G, \top \rangle$, **then**
6:    **return true;**
7: **else if** expanded_graph = $\langle G, \lambda' \rangle$ and $\lambda'(A) = 1$, **then**
8:    **return true;**
9: **else**
10:    **return false.**
11: **end if**

Algorithm A.3.3 Expand($\langle G, \lambda \rangle$)

1: push all formulae A such that $\lambda(A) \neq \bot$ into formula_stack;
2: **while** $\lambda \neq \top$ and formula_stack is not empty **do**
3:    pop a formula A from formula_stack
4:    Apply_Intelim($A, \langle G, \lambda \rangle$)
5: **end while**

Algorithm A.3.4 Apply_Intelim($A, \langle G, \lambda \rangle$)

1: **for all** the G-modules M containing A **do**
2:    let B be the top formula of M;
3:    consider the set $\Delta = \{T\,C \mid C \in M \text{ and } \lambda(C) = 1\} \cup \{F\,C \mid C \in M \text{ and } \lambda(C) = 0\}$;
4:    **if** Δ contains all the premises for an application of an introduction rule with conclusion $S\,B$ **then**
5:      **if** $\lambda(B) = |1 - \mathtt{val}(S)|$ **then**
6:         **return** $\lambda = \top$
7:      **else if** $\lambda(B) = \bot$ **then**
8:         set $\lambda(B) = \mathtt{val}(S)$ and push B into `formula_stack`;
9:      **else**
10:         do nothing;
11:      **end if**
12:    **else**
13:      **for all** $C \in M$ **do**
14:         **if** Δ contains all the premises for an application of an elimination rule with major premise $\mathtt{sign}(\lambda(B))\,B$ and conclusion $S\,C$, **then**
15:           **if** $\lambda(C) = |1 - \mathtt{val}(S)|$ **then**
16:             **return** $\lambda = \top$
17:           **else if** $\lambda(C) = \bot$ **then**
18:             set $\lambda(C) = \mathtt{val}(S)$ and push C into `formula_stack`;
19:           **else**
20:             do nothing;
21:           **end if**
22:         **end if**
23:      **end for**
24:    **end if**
25: **end for**

Note that:

1. the number of nodes in the subformula graph G (line 1 of both decision procedures) is $O(n)$;

2. the cost of building the initial labelled subformula graph $\langle G, \lambda \rangle$ is $O(n^2)$;

3. the **while** loop in the Expand subroutine (Algorithm A.3.3) is executed at most as many times as there are nodes in G, namely $O(n)$ times; (here the key observation is that in Line 3 the formula A can be safely removed from formula_stack, that is each formula in formula_stack needs to be visited at most once, see p. 58;)

4. the essential cost of each run of the **while** loop consists in the cost of the Apply_Intelim subroutine (Algorithm A.3.4);

5. for each formula A in G there are at most $O(n)$ G-modules containing A (line 1 of Algorithm A.3.4);

6. the maximum number of nodes in a G-module is $a+1$, where a is the maximum arity of a logical operator in $\mathcal{L}$;

7. the cost of each run of the **forall** loop in the Apply_Intelim subroutine is $O(a)$.

A.4 Proof of Proposition 2.1.20

Proof. First, suppose $\mathcal{T}$ is a proof of φ. By Clause (i) of Definition 2.1.16, ψ_n cannot be idle, and so it is a premise of an application of some rule. Given that the sequence of introductions is maximal in the branch, ψ_n is not used as premise of an introduction. Moreover, by Lemma 1.4.11, non redundant proofs contain no detours, and so ψ_n is not used as major premise of an elimination. Thus either $\psi_n = \varphi$, or ψ_n is used as minor premise of an elimination, or ψ_n is used as premise of an application of the closure rule.

If $\mathcal{T}$ is a refutation of X, again ψ_n is not used as premise of an introduction and cannot be a detour. Hence, either it is used as minor premise of an elimination or as premise an application of the closure rule. □

A.5 Proof of Lemma 2.1.17

Proof. We use the notation 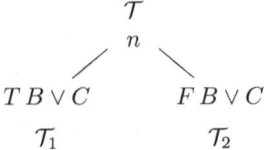 to denote either an empty intelim tree or a non-empty intelim tree such that n is one of its terminal nodes. The proof is by lexicographic induction on $\langle \gamma(\mathcal{T}), \lambda(\mathcal{T})\rangle$, where $\gamma(\mathcal{T})$ is the maximum logical complexity[1] of a PB-formula in $\mathcal{T}$ that is not g-analytic and $\lambda(\mathcal{T})$ is the number of occurrences of such non-f-analytic PB-formulae of maximal complexity.

Let $\gamma(\mathcal{T}) = m > 0$ and let A be a PB-formula of logical complexity m. There are several cases depending on the logical form of A. We discuss only the case $A = B \vee C$, the other cases being similar.

If $A = B \vee C$, then $\mathcal{T}$ has the following form:

$$\begin{array}{c} \mathcal{T} \\ n \\ \swarrow \quad \searrow \\ TB\vee C \qquad FB\vee C \\ \mathcal{T}_1 \qquad\qquad \mathcal{T}_2 \end{array}$$

where $\mathcal{T}_1$ and $\mathcal{T}_2$ are intelim trees such that each of their open branches contains φ (or are both closed intelim trees in case $\mathcal{T}$ is a refutation of X). Let $\mathcal{T}'$ be the following intelim tree:

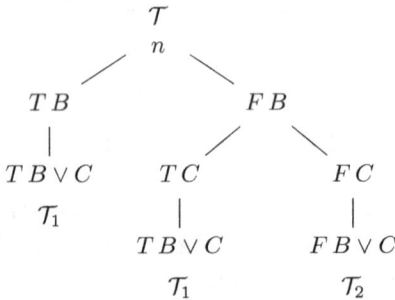

Clearly $\mathcal{T}'$ is a $k+1$-depth intelim proof of φ from X (a $k+1$-depth intelim refutation of X). Moreover, either $\gamma(\mathcal{T}') < \gamma(\mathcal{T})$, or $\gamma(\mathcal{T}') = \gamma(\mathcal{T})$ and $\lambda(\mathcal{T}') < \lambda(\mathcal{T})$. □

---

[1] The logical complexity of a formula is the number of occurrences of logical operators in it.

A.6 The full expand algorithm

Algorithm A.6.1 is a generalization for $k \geq 0$ of the Expand sub-routine of the algorithm in Section A.3. In the main algorithm (A.3.2) line 4 should be replaced by
4: set expanded_graph = Depth-Expand($k, \langle G, \lambda \rangle$).

Algorithm A.6.1 Depth-Expand($k, \langle G, \lambda \rangle$)

1: **if** $k = 0$, **then**
2:    Expand($\langle G, \lambda \rangle$);
3: **else**
4:    push all formulae A such that $\lambda(A) = \bot$ into undefined_formulae;
5:    **while** $\lambda \neq \top$, $\lambda(A) \neq 1$ and undefined_formulae is not empty **do**
6:      pop a formula A from undefined_formulae;
7:      let $\lambda_1(A) = 1$ and $\lambda_1(B) = \lambda(B)$ for all $B \neq A$;
8:      let $\lambda_2(A) = 0$ and $\lambda_2(B) = \lambda(B)$ for all $B \neq A$;
9:      Depth-Expand($k - 1, \langle G, \lambda_1 \rangle$);
10:     Depth-Expand($k - 1, \langle G, \lambda_2 \rangle$);
11:     **if** $\lambda_1 = \top$, **then**
12:        set $\lambda = \lambda_2$;
13:        remove from undefined_formulae the formulae A such that $\lambda(A) \neq \bot$;
14:     **else**
15:        **if** $\lambda_2 = \top$, **then**
16:           set $\lambda = \lambda_1$,
17:           remove from undefined_formulae the formulae A such that $\lambda(A) \neq \bot$;
18:        **else**
19:           set $\lambda(A) = x$ for all A such that $\lambda_1(A) = \lambda_2(A)$
20:           remove from undefined_formulae the formulae A such that $\lambda(A) \neq \bot$;
21:        **end if**
22:     **end if**
23:    **end while**
24: **end if**

Bibliography

AMGOUD, L., & CAYROL, C. 2002. Inferring from inconsistency in preference-based argumentation frameworks. *International Journal of Automated Reasoning*, **29**, 125–169.

ANDERSON, A. R., & BELNAP JR, N.D. 1975. *Entailment: the Logic of Relevance and Necessity*. Vol. 1. Princeton: Princeton University Press.

ANDERSON, R.L. 2015. *The Poverty of Conceptual Truth: Kant's Analytic/Synthetic Distinction and the Limits of Metaphysics*. Oxford: Oxford University Press.

ARTEMOV, S., & KUZNETS, R. 2009. Logical omniscience as a computational complexity problem. *Pages 14–23 of:* HEIFETZ, A. (ed), *Proceedings of the Twelfth Conference on Theoretical Aspects of Rationality and Knowledge (TARK '09)*. New York: ACM.

ARTEMOV, S., & KUZNETS, R. 2014. Logical omniscience as infeasibility. *Annals of Pure and Applied Logic*, **165**(1), 6–25. The Constructive in Logic and Applications.

ASCHIERI, F., CIABATTONI, A., & GENCO, F.A. 2018. Classical proofs as parallel programs. *Pages 43–57 of:* ORLANDINI, A., & ZIMMERMANN, M. (eds), *Proceedings of the Ninth Symposium on Games, Automata, Logics and Formal Verification (GandALF '18)*, vol. 277.

AVRON, A. 1991. Natural 3-valued Logics. Characterization and Proof Theory. *The Journal of Symbolic Logic*, **56**(1), 276–294.

AVRON, A. 2005a. Non-deterministic matrices and modular semantics of rules. *Pages 155–174 of:* BEZIAU, J.-Y. (ed), *Logica Universalis*. Birkäuser.

AVRON, A. 2005b. A non-deterministic view on non-classical negations. *Studia Logica*, **80**(2), 159–194.

AVRON, A., & LEV, I. 2001. Canonical propositional Gentzen-type systems. *Pages 529–544 of:* GORE, R., LEITSCH, A., & NIPKOV, T. (eds), *Automated Reasoning: First Joint International Conference (IJCAR '01)*. Lecture Notes in Artificial Intelligence, vol. 2083. Springer.

AVRON, A., & ZAMANSKY, A. 2006. Cut elimination and quantification in canonical systems. *Studia Logica*, **82**, 157–176.

AYER, A.J. 1958. *Language, Truth and Logic*. London: V. Gollancs.

BALDI, P., & HOSNI, H. 2020. Depth-bounded belief functions. *International Journal of Approximate Reasoning*, **123**, 26–40.

BALDI, P., & HOSNI, H. 2021 (06–09 Jul). Logical approximations of qualitative probability. *Pages 12–21 of:* CANO, A., DE BOCK, J., MIRANDA, E., & MORAL, S. (eds), *Proceedings of the Twelfth International Symposium on Imprecise Probability: Theories and Applications*. Proceedings of Machine Learning Research (PMLR), vol. 147.

BALDI, P., & HOSNI, H. 2023. A logic-based tractable approximation of probability. *Journal of Logic and Computation*, **33**, 599–622.

BALDI, P., D'AGOSTINO, M., & HOSNI, H. 2020. Depth-bounded approximations of probability. *Pages 607–621 of:* LESOT, M.-J., VIEIRA, S., REFORMAT, M.Z., CARVALHO, J.P., WILBIK, A., BOUCHON-MEUNIER, B., & YAGER, R.R. (eds), *Information Processing and Management of Uncertainty in Knowledge-Based Systems*. Cham: Springer International Publishing.

BALDI, P., D'ASARO, F.A., , & PRIMIERO, G. 2021. Introducing k-lingo: A k-depth bounded version of ASP system clingo. *Pages 661–664 of:* BIENVENU, M., LAKEMEYER, G., & ERDEM, E. (eds), *Proceedings of the Eighteenth International Conference on Principles of Knowledge Representation and Reasoning (KR '21)*. JCAI Organization.

BAR-HILLEL, Y., & CARNAP, R. 1953. Semantic information. *British Journal for the Philosophy of Science*, **4**(14), 147–157.

BAR-HILLEL, Y., & CARNAP, R. 1964. An outline of a theory of semantic information [1953]. *Pages 221–74 of: Language and Information: Selected Essays on Their Theory and Application*. Reading, Massachusetts and London: Addison-Wesley.

BELNAP JR., N. D. 1976. How a computer should think. *Pages 30–55 of:* RYLE, G. (ed), *Contemporary Aspects of Philosophy*. Oriel Press.

BELNAP JR., N. D. 1977. A useful four-valued logic. *Pages 8–37 of:* DUNN, J. M., & EPSTEIN, G. (eds), *Modern uses of multiple-valued logics*. Dordrecht: Reidel.

BENDALL, K. 1978. Natural deduction, separation and the meaning of logical operators. *Journal of Philosophical Logic*, **7**, 245–276.

BENNETT, J. 1969. Entailment. *The Philosophical Review*, **78**, 197–236.

BERGSTRÖM, L., & FØLLESDAG, D. 1994. Interview with Willard van Orman Quine, November 1993. *Theoria*, **60**, 193–206.

BESOLD, T.R., D'AVILA GARCEZ, A., BADER, S., BOWMAN, H., DOMINGOS, P., HITZLER, P., KUEHNBERGER, K., LAMB, L.C., LOWD, D., LIMA, P.M.V., DE PENNING, L., PINKAS, G., POON, H., & ZAVERUCHA, G. 2017. *Neural-symbolic learning and reasoning: A survey and interpretation*. arXiv:1711.03902v1.

BEZOU-VRAKATSELI, E., COCARASCU, O., & MODGIL, S. 2024. *Towards Dialogues for Joint Human-AI Reasoning and Value Alignment*.

BJERRING, J.C., & SKIPPER, M. 2019. A dynamic solution to the problem of logical omniscience. *Journal of Philosophical Logic*, **48**(3), 501–521.

BJÖRK, M. 2003. *Stålmarck's Method for Automated Theorem Proving in First Order Logic*. Ph.D. thesis, Göteborg University.

BJÖRK, M. 2005. A first-order extension of Stålmarck's method. *Pages 276–291 of:* SUTCLIFFE, G., & VORONKOV, A. (eds), *LPAR 2005.* LNAI 3835. Springer.

BJÖRK, M. 2009. First-order Stålmarck. Universal lemmas through branch merges. *Journal of Automated Reasoning*, **42**, 99–122.

BOLZANO, B. 2014. *Theory of Science [1837].* Oxford: Oxford University Press.

BREWKA, G. 1989. Preferred subtheories: An extended logical framework for default reasoning. *Pages 1043–1048 of: Proceedings of the Eleventh International Joint Conference on Artificial Intelligence (IJCAI '89).* San Francisco: Morgan Kaufmann Publishers Inc.

BREWKA, G. 1994. Adding priorities and specificity to default logic. *Pages 247–260 of:* MACNISH, C., PEARCE, D., & PEREIRA, L. MONIZ (eds), *European Workshop on Logics in Artificial Intelligence (JELIA '94).* Springer.

CADOLI, M., & SCHAERF, M. 1992. Approximate reasoning and non-omniscient agents. *Pages 169–183 of: Proceedings of the Fourth Conference on Theoretical Aspects of Reasoning about Knowledge (TARK '92).* San Francisco: Morgan Kaufmann Publishers Inc.

CADOLI, M., & SCHAERF, M. 1993. A survey of complexity results for non-monotonic logics. *Journal of Logic Programming*, **17**, 127–160.

CAMINADA, M., & AMGOUD, L. 2007. On the evaluation of argumentation formalisms. *Artificial Intelligence*, **171**(5-6), 286–310.

CAMINADA, M., CARNIELLI, W., & DUNNE, P. 2012. Semi-stable semantics. *Journal of Logic and Computation*, **22**(5), 1207–1254.

CARAPEZZA, M., & D'AGOSTINO, M. 2010. Logic and the myth of the perfect language. *Logic and Philosophy of Science*, **8**(1), 1–29.

CARNAP, R. 1952. Meaning Postulates. *Philosophical Studies*, **3**(5), 65–73.

CARNAP, R., HAHN, H., & NEURATH, O. 1973. The scientific conception of the world [1929]. *In:* NEURATH, M., & COHEN, R.S. (eds), *Empiricism and Sociology*. Dordrecht: Reidel.

CHERNIAK, C. 1986a. *Minimal Rationality*. MIT Press.

CHERNIAK, C. 1986b. *Minimal Rationality*. MIT Press.

CIGNARALE, G., & PRIMIERO, G. 2021. A multi-agent depth bounded Boolean logic. *Pages 176–191 of:* CLEOPHAS, L., & MASSINK, M. (eds), *Software Engineering and Formal Methods (SEFM '20)*. Cham: Springer International Publishing.

COHEN, M.R., & NAGEL, E. 1934. *An Introduction to Logic and Scientific Method*. New York: Harcourt.

CONAN DOYLE, A. 1981. *The Penguin Complete Sherlock Holmes*. London: Penguin Books.

COOK, S. A., & RECKHOW, R. 1974. On the length of proofs in the propositional calculus. *Pages 135–148 of:* CONSTABLE, R.L., RITCHIE, R.W., CARLYLE, J.W., & HARRISON, M.A. (eds), *Proceedings of the Sixth Annual Symposium on the Theory of Computing (STOC '74)*. New York: Association for Computing Machinery.

COOK, S.A. 1971. The complexity of theorem-proving procedures. *Pages 151–158 of: Proceedings of the Third Annual ACM Symposium on Theory of Computing (STOC '71)*. New York: ACM.

CRAWFORD, J.M., & ETHERINGTON, D.W. 1998. A non-deterministic semantics for tractable inference. *Pages 286–291 of: Proceedings of the AAAI Conference on Artificial Intelligence (AAAI '98)*. AAAI Press.

D'AGOSTINO, M. 1990. *Investigations Into the Complexity of Some Propositional Calculi*. PRG Technical Monographs 88. Oxford University Computing Laboratory.

D'AGOSTINO, M. 1992. Are tableaux an improvement on truth tables? Cut-free proofs and bivalence. *Journal of Logic, Language and Information*, **1**, 235–252.

D'AGOSTINO, M. 1999. Tableau methods for classical propositional logic. *Pages 45–123 of:* D'AGOSTINO, M., GABBAY, D.M., HÄHNLE, R., & POSEGGA, J. (eds), *Handbook of Tableaux Methods*. Kluwer Academic Publishers.

D'AGOSTINO, M. 2005. Classical natural deduction. *Pages 429–468 of:* ARTËMOV, S.N., BARRINGER, H., D'AVILA GARCEZ, A., LAMB, L.C., & WOODS, J. (eds), *We will show them! Essays in Honour of Dov Gabbay*, vol. 1. College Publications.

D'AGOSTINO, M. 2010. Tractable depth-bounded logics and the problem of logical omniscience. *Pages 245–275 of:* MONTAGNA, F., & HOSNI, H. (eds), *Probability, Uncertainty and Rationality*. CRM series. Springer.

D'AGOSTINO, M. 2013a. Depth-bounded logic for realistic agents. *L&PS – Logic and Philosophy of Science*, **11**, 3–57.

D'AGOSTINO, M. 2013b. Semantic information and the trivialization of logic: Floridi on the scandal of deduction. *Information*, **4**(1), 33–59.

D'AGOSTINO, M. 2014a. Analytic inference and the informational meaning of the logical operators. *Logique et Analyse*, **227**, 407–437.

D'AGOSTINO, M. 2014b. Informational semantics, non-deterministic matrices and feasible deduction. *Electronic Notes in Theoretical Computer Science*, **305**, 35–52.

D'AGOSTINO, M. 2015. An informational view of classical logic. *Theoretical Computer Science*, **606**, 79–97.

D'AGOSTINO, M. 2016. The philosophy of mathematical information. *Pages 164–179 of:* FLORIDI, L. (ed), *Routledge Handbook of the Philosophy of Information*. London: Routledge.

D'AGOSTINO, M. 2019. An informational approach to feasible deduction. *Pages 89–115 of:* BELLOTTI, L., GILI, L., MORICONI, E., & TURBANTI, G. (eds), *Third Pisa Colloquium in Logic, Language and Epistemology: Essays in Honour of Mauro Mariani and Carlo Marletti*. Edizioni ETS.

D'AGOSTINO, M., & FLORIDI, L. 2009. The enduring scandal of deduction: Is propositional logic really uninformative? *Synthese*, **167**(2), 271–315.

D'AGOSTINO, M., & FLORIDI, L. 2015. Towards a more realistic theory of semantic information. *Pages 221–227 of:* BEZIAU, J-Y., & BUCHSBAUM, A. (eds), *Handbook of the Fifth World Congress and School on Universal Logic (UNILOG '15)*. University of Instanbul, Turkey.

D'AGOSTINO, M., & MODGIL, S. 2016. A rational account of classical logic argumentation for real-world agents. *Pages 141–149 of: Proceedings of European Conference on Artificial Intelligence (ECAI '16)*. IOS Press.

D'AGOSTINO, M., & MODGIL, S. 2018a. Classical logic, argument and dialectic. *Artificial Intelligence*, **262**, 15–51.

D'AGOSTINO, M., & MODGIL, S. 2018b. A study of argumentative characterisations of preferred subtheories. *Pages 1788–1794 of:* LANG, J. (ed), *Proceedings of the Twenty-Seventh International Joint Conference on Artificial Intelligence (IJCAI '18)*. AAAI Press.

D'AGOSTINO, M., & MODGIL, S. 2020. A fully rational account of structured argumentation under resource bounds. *Pages 1841–1847 of:* BESSIERE, C. (ed), *Proceedings of the Twenty-Ninth International Joint Conference on Artificial Intelligence (IJCAI '20)*. International Joint Conferences on Artificial Intelligence Organization.

D'AGOSTINO, M., & MONDADORI, M. 1994. The taming of the cut. *Journal of Logic and Computation*, **4**, 285–319.

D'AGOSTINO, M., & SOLARES-ROJAS, A. 2022. Towards tractable approximations to many-valued logics: The case of first degree entailment. *Pages 57–76 of:* SEDLÁR, I. (ed), *The Logica Yearbook 2021*. College Publications.

D'AGOSTINO, M., & SOLARES-ROJAS, A. 2024. Tractable depth-bounded approximations to FDE and its satellites. *Journal of Logic and Computation*, **34**, 815–855.

D'AGOSTINO, M., BRODA, K., & MONDADORI, M. 2006. A solution to a problem of Popper. *Pages 147–168 of:* ALAI, M., & TAROZZI, G. (eds), *Popper Philosopher of Science*. Rubbettino.

D'AGOSTINO, M., FINGER, M., & GABBAY, D.M. 2013. Semantics and proof-theory of depth-bounded Boolean logics. *Theoretical Computer Science*, **480**, 43–68.

D'AGOSTINO, M., GABBAY, D., & MODGIL, S. 2020. Normality, non-contamination and logical depth in classical natural deduction. *Studia Logica*, **108**(2), 291–357.

D'AGOSTINO, M., LARESE, C., & MODGIL, S. 2021. Towards depth-bounded natural deduction for classical first-order logic. *Journal of Applied Logics – IfCoLog*, **8**(2), 423–451.

DALAL, M. 1996. Anytime families of tractable propositional reasoners. *Pages 42–45 of: Proceedings of the Fourth International Symposium on AI and Mathematics (AI/MATH '96)*.

DALAL, M. 1998. Anytime families of tractable propositional reasoners. *Annals of Mathematics and Artificial Intelligence*, **22**, 297–318.

DAVEY, B.A., & PRIESTLEY, H.A. 2002. *Introduction to Lattices and Order*. Cambridge University Press.

DE JONG, W.R. 1995. Kant's analytic judgments and the traditional theory of concepts. *Journal of the History of Philosophy*, **33**, 631–641.

DECOCK, L. 2006. *True by virtue of meaning. Carnap and Quine on some analytic-synthetic distinctions.* https://www.academia.edu/837919/Carnap_and_Quine_on_some_analytic_synthetic_distinctions.

DUMMETT, M. 1977. *Elements of Intuitionism*. Oxford: Clarendon Press.

DUMMETT, M. 1978. The philosophical basis of intuitionistic logic. *In: Truth and Other Enigmas*. Cambridge University Press.

DUMMETT, M. 1991a. *Frege: Philosophy of Mathematics*. London: Duckworth.

DUMMETT, M. 1991b. *The Logical Basis of Metaphysics*. London: Duckworth.

DUNG, P.M. 1995. On the acceptability of arguments and its fundamental role in nonmonotonic reasoning, logic programming and n-person games. *Artificial Intelligence*, **77**(2), 321–358.

DUNG, P.M., TONI, F., & MANCARELLA, P. 2010. Some design guidelines for practical argumentation systems. *Pages 183–194 of:* BARONI, P., CERUTTI, F., GIACOMIN, M., & SIMARI, G.R. (eds), *Proceedings of the Conference on Computational Models of Argument (COMMA '10)*. Amsterdam: IOS Press.

ÉGRÉ, P. 2020. Logical omniscience. *Pages 1–25 of: The Wiley Blackwell Companion to Semantics*. John Wiley & Sons, Ltd.

EITER, T., & GOTTLOB, G. 1995. The complexity of logic-based abduction. *Journal of the ACM*, **2**(1), 3–42.

FAGIN, R., HALPERN, J.Y., & VARDI, M.Y. 1995a. A nonstandard approach to the logical omniscience problem. *Artificial Intelligence*, **79**, 203–240.

FAGIN, R., HALPERN, J. Y., MOSES, Y., & VARDI, M. Y. 1995b. *Reasoning About Knowledge*. Cambridge, MA, USA: MIT Press.

FAN, X., & TONI, F. 2014. A general framework for sound assumption-based argumentation dialogues. *Artificial Intelligence*, **216**, 20–54.

FERREIRA, F. 2008. The coordination principle: A problem for bilateralism. *Mind*, **117**, 1051–1057.

FINGER, M. 2004a. Polynomial approximations of full propositional logic via limited bivalence. *Pages 526–538 of: Ninth European Conference on Logics in Artificial Intelligence (JELIA '04)*. Lecture Notes in Artificial Intelligence, vol. 3229. Springer.

FINGER, M. 2004b. Towards polynomial approximations of full propositional logic. *Pages 11–20 of:* BAZZAN, A.L.C., & LABIDI, S. (eds),

XVII Brazilian Symposium on Artificial Intelligence (SBIA '04). Lecture Notes in Artificial Intellingence, vol. 3171. Springer.

FINGER, M., & GABBAY, D.M. 2006. Cut and pay. *Journal of Logic, Language and Information*, **15**(3), 195–218.

FINGER, M., & WASSERMANN, R. 2004. Approximate and limited reasoning: Semantics, proof theory, expressivity and control. *Journal of Logic and Computation*, **14**(2), 179–204.

FINGER, M., & WASSERMANN, R. 2006. The universe of propositional approximations. *Theoretical Computer Science*, **355**(2), 153–166.

FLORIDI, L. 2004. Outline of a theory of strongly semantic information. *Minds and Machines*, **14**(2), 197–222.

FLORIDI, L. 2006. The logic of being informed. *Logique et Analyse*, **49**(196), 433–460.

FLORIDI, L. 2011. *The Philosophy of Information*. Oxford: Oxford University Press.

FREGE, G. 1960. *The Foundations of Arithmetic [1884]*. Evanston, Illinois: Northwestern University Press.

FREGE, G. 1979. *Posthumous Writings*. Oxford: Basil Blackwell.

FREGE, G. 1984. *Collected Papers on Mathematics, Logic, and Philosophy*. Oxford: Basil Blackwell.

FRISCH, A.M. 1987. Inference without chaining. *Pages 515–519 of: Proceedings of the Tenth International Joint Conference on Artificial Intelligence (IJCAI '87)*. San Francisco, CA, USA: Morgan Kaufmann Publishers Inc.

GABBAY, D.M. 1991. What is a logical system? *Pages 179–216 of:* GABBAY, D.M. (ed), *What Is a Logical System?* Clarendon Press.

GABBAY, D.M., & OWENS, R. 1991. Temporal logics for real-time systems. *Pages 97–103 of: Proceedings of the IMACS Symposium on the Modelling and Control of Technological Systems (MCTS '91)*, vol. 2.

GABBAY, D.M., & WOODS, J. 2001. The new logic. *Logic Journal of the IGPL*, **9**(2), 141–174.

GABBAY, DOV M. 2014. What is a logical system? An evolutionary view: 1964–2014. *Pages 41–132 of:* SIEKMANN, J. H. (ed), *Computational Logic*. Handbook of the History of Logic, vol. 9. North-Holland.

GABBAY, M. 2017. Bilateralism does not provide a proof theoretic treatment of classical logic (for technical reasons). *Journal of Applied Logic*, **5**, 108–122.

GENTZEN, G. 1969. Investigations into logical deduction [1935]. *Pages 68–131 of:* SZABO, M. (ed), *The Collected Papers of Gerhard Gentzen*. Amsterdam: North-Holland.

GOROGIANNIS, N., & HUNTER, A. 2011. Instantiating abstract argumentation with classical logic arguments: Postulates and properties. *Artificial Intelligence*, **175**, 1479–1497.

HADFIELD-MENELL, D., DRAGAN, A., ABBEEL, P., & RUSSELL, S. 2016. Cooperative inverse reinforcement learning. *Pages 3916–3924 of:* LEE, D.D., VON LUXBURG, U., GARNETT, R., SUGIYAMA, M., & GUYON, I. (eds), *Proceedings of the Thirtieth International Conference on Neural Information Processing Systems (NIPS '16)*. Red Hook, NY: Curran Associates Inc.

HAHN, H. 1959. Logic, mathematics and knowledge of nature. *Pages 147–161 of:* AYER, A.J. (ed), *Logical Positivism*. Glencoe, Illinois: The Free Press.

HÄHNLE, R. 2001. Tableaux and related methods. *Pages 101–178 of:* ROBINSON, A., & VORONKOV, A. (eds), *Handbook of Automated Reasoning*, vol. 1. Amsterdam: North Holland.

HALPERN, J.Y. 1995. Reasoning about knowledge: A survey. *Pages 1–34 of:* GABBAY, D.M., HOGGER, C.J., & ROBINSON, J.A. (eds), *Handbook of Logic in Artificial Intelligence and Logic Programming*, vol. 4. Oxford: Clarendon Press.

HANNA, R. 2001. *Kant and the Foundations of Analytic Philosophy.* Oxford: Clarendon Press.

HAWKE, P., ÖZGÜN, A., & BERTO, F. 2020a. The Fundamental Problem of Logical Omniscience. *Journal of Philosophical Logic*, **49**(4), 727–766.

HAWKE, P., ÖZGÜN, A., & BERTO, F. 2020b. The fundamental problem of logical omniscience. *Journal of Philosophical Logic*, **49**(4), 727–766.

HEMPEL, C.G. 2001. *The Philosophy of Carl G. Hempel. Studies in Science, Explanation, and Rationality.* Oxford: Oxford University Press.

HINTIKKA, J. 1973. *Logic, Language Games and Information: Kantian Themes in the Philosophy of Logic.* Oxford: Clarendon Press.

HUMBERSTONE, L. 2000. The revival of rejective negation. *Journal of Philosophical Logic*, **29**, 331–381.

INDRZEJCZAK, A. 2010. *Natural Deduction, Hybrid Systems and Modal Logics.* Dordrecht: Springer.

KANOVICH, M.I. 1992. Horn programming in linear logic is NP-complete. *Pages 200–210 of: Proceedings of the Seventh Annual Symposium on Logic in Computer Science.* IEEE Computer Science Press.

KANT, I. 1997. *Prolegomena to Any Future Metaphysics that May Be Able to Come forward as a Science [1783].* Cambridge: Cambridge University Press.

KANT, I. 1998. *Critique of Pure Reason [1781,1787].* Cambridge: Cambridge University Press.

KLEENE, S.C. 1952. *Introduction to Metamathematics.* Amsterdam: North-Holland.

KLEENE, S.C. 1967. *Mathematical Logic.* New York: John Wiley & Sons, Inc.

KRAUS, S., LEHMANN, D., & MAGIDOR, M. 1990. Nonmonotonic reasoning, preferential models and cumulative logics. *Artificial Intelligence*, **44**(1), 167–207.

KRIPKE, S.A. 1965. Semantical analysis of intuitionistic logic I. *Pages 92–130 of:* CROSSLEY, J., & DUMMETT, M.A.E. (eds), *Formal Systems and Recursive Functions*. Amsterdam: North-Holland Publishing.

LADNER, R.E. 1977. The computational complexity of provability in systems of model propositional logic. *SIAM Journal of Computing*, **6**(3), 467–480.

LAKEMEYER, G., & LEVESQUE, H.J. 2020. A first-order logic of limited belief based on possible worlds. *Pages 624–635 of:* CALVANESE, D., ERDEM, E., & THIELSCHER, M. (eds), *Proceedings of the Seventeenth International Conference on Principles of Knowledge Representation and Reasoning (KR '20)*. IJCAI Organization.

LAMPERT, T. 2017. Minimizing disjunctive normal forms of pure first-order logic. *Logic Journal of the IGPL*, **25**(3), 325–347.

LANG, J. 2015. Twenty-five years of preferred subtheories. *Pages 157–172 of:* EITER, T., STRASS, H., TRUSZCZYŃSKI, M., & WOLTRAN, S. (eds), *Advances in Knowledge Representation, Logic Programming, and Abstract Argumentation: Essays Dedicated to Gerhard Brewka on the Occasion of His Sixtieth Birthday*. Springer.

LANGFORD, C.H. 1992. The notion of analysis in Moore's philosophy. *Pages 321–342 of:* SCHLIPP, P.A. (ed), *The Philosophy of G.E. Moore*. LaSalle, Illinois: Open Court.

LARESE, C. 2019. *The Principle of Analyticity of Logic: A Philosophical and Formal Perspective*. Ph.D. thesis, Scuola Normale Superiore di Pisa.

LARESE, C. 2020. Notes on Hintikka's analytic-synthetic distinction. *Isonomia – Epistemologica*, 1–31.

LARESE, C. 2022. Kant on the analyticity of logic. *Argumenta*, **8**, 173–187.

LARESE, C. 2023a. Hintikka's conception of syntheticity as the introduction of new individual. *Synthese*, **201**.

LARESE, C. 2023b. Hintikka's conception of syntheticity as the introduction of new individuals. *Synthese*, **201**.

LAROTONDA, M., & PRIMIERO, G. 2023. A depth-bounded semantics for becoming informed. *Pages 366–382 of:* MASCI, P., BERNARDESCHI, C., GRAZIANI, P., KODDENBROCK, M., & PALMIERI, M. (eds), *Software Engineering and Formal Methods (SEFM '22)*. Cham: Springer International Publishing.

LEITGEB, H., & CARUS, A. 2024. Rudolf Carnap. *In:* ZALTA, EDWARD N., & NODELMAN, URI (eds), *The Stanford Encyclopedia of Philosophy*, Fall 2024 edn. Metaphysics Research Lab, Stanford University.

LEVESQUE, H.J. 1984. A Logic of Implicit and Explicit Belief. *In: AAAI Conference on Artificial Intelligence.*

LEVESQUE, H.J. 1988. Logic and the complexity of reasoning. *Journal of Philosophical Logic*, **17**(4), 355–389.

LINCOLN, P. 1995. Deciding probability of linear logic formulas. *Pages 197–210 of:* GIRARD, J.-Y, LAFONT, Y., & REGNIER, L. (eds), *Proceedings of the Workshop on Advances in Linear Logic*. Cambridge University Press.

LINCOLN, P., & WINKLER, T. 1994. Constant-Only Multiplicative Linear Logic is NP-Complete. *Theoretical Computer Science*, **135**, 155–169.

LINCOLN, P., MITCHELL, J., SCEDROV, A., & SHANKAR, N. 1992. Decision problems for propositional linear logic. *Annals of pure and applied logic*, **56**, 239–311.

MAKINSON, D. 2013. On an inferential semantics for classical logic. *Logic Journal of IGPL*, **22**, 147–154.

MANCOSU, P., GALVAN, S., & ZACH, R. 2021. *An Introduction to Proof Theory: Normalization, Cut-Elimination, and Consistency Proofs*. Oxford: Oxford University Press.

MARCUS, G.F. 2018. *Deep learning: A critical appraisal.* `arXiv:1801.00631v1`.

MAREK, W., & TRUSZCZYSKI, M. 1991. Autoepistemic logic. *Journal of the ACM*, **38**(3), 587–618.

MASSACCI, F. 1998. *Efficient Approximate Deduction and an Application to Computer Security*. Ph.D. thesis, Università degli Studi di Roma "La Sapienza".

MCCARTHY, J. 1986. Applications of circumscription to formalizing common-sense knowledge. *Artificial Intelligence*, **28**(1), 89–116.

MCDERMOTT, D. 1982. Nonmonotonic logic II: Nonmonotonic modal theories. *Journal of the ACM*, **29**(1), 33–57.

MENDONÇA, B.R. 2023. Dialogue games and deductive information: a dialogical account of the concept of virtual information. *Synthese*, **202**(3), 73.

MEYER, J.C. 2003. Modal epistemic and doxastic logic. *Pages 1–38 of:* GABBAY, D.M., & GUENTHNER, F. (eds), *Handbook of Philosophical Logic*, 2nd edn., vol. 10. Kluwer Academic Publishers.

MODGIL, S. 2017a. Dialogical scaffolding for human and artificial agent reasoning. *Pages 58–71 of:* DIAKIDOY, I.-A., KAKAS, A.C., LIETO, A., & MICHAEL, L. (eds), *Proceedings of the Fifth International Workshop on AI and Cognition (AIC '17)*, vol. 2090. CEUR.

MODGIL, S. 2017b. Towards a general framework for dialogues that accommodate reasoning about preferences. *Pages 175–191 of:* MODGIL, S., OREN, N., & TONI, F. (eds), *Theory and Applications of Formal Argumentation*. Springer.

MODGIL, S. 2018. Many kinds of minds are better than one: Value alignment through dialogue. *In: Workshop on Argumentation and Philosophy (co-located with COMMA '18)*.

MODGIL, S., & CAMINADA, M. 2009. Proof theories and algorithms for abstract argumentation frameworks. *Pages 105–129 of:* RAHWAN, I., & SIMARI, G. (eds), *Argumentation in AI*. Springer.

MODGIL, S., & PRAKKEN, H. 2013. A general account of argumentation and preferences. *Artificial Intelligence*, **195**(0), 361–397.

MONDADORI, M. 1988a. *Classical analytical deduction.* Annali dell'Università di Ferrara, Sez. III, Discussion paper 1. Università di Ferrara.

MONDADORI, M. 1988b. *Classical analytical deduction: Part II.* Annali dell'Università di Ferrara, Sez. III, Discussion paper 5. Università di Ferrara.

MONDADORI, M. 1988c. On the notion of a classical proof. *Pages 211–224 of: Temi e prospettive della logica e della filosofia della scienza contemporanee,* vol. 1. Bologna: CLUEB.

MONDADORI, M. 1988d. On the notion of a classical proof. *Pages 211–224 of: Temi e prospettive della logica e della filosofia della scienza contemporanee,* vol. 1. Bologna: CLUEB.

MONDADORI, M. 1989a. *An improvement of Jeffrey's deductive trees.* Annali dell'Università di Ferrara, Sez. III, Discussion paper 7. Università di Ferrara.

MONDADORI, M. 1989b. *An Improvement of Jeffrey's Deductive Trees.* Annali dell'Università di Ferrara; Sez. III; Discussion paper 7. Università di Ferrara.

MONDADORI, M. 1995. Efficient inverse tableaux. *Logic Journal of the IGPL,* **3**(6), 939–953.

MUGNAI, M. 2016. Ontology and logic: The case of scholastic and late-scholastic theory of relations. *British Journal for the History of Philosophy,* **24**(3), 532–553.

NEGRI, S., & VON PLATO, J. 2001. *Structural Proof Theory.* New York: Cambridge University Press.

NELTE, K. 1997. *Formulas of First-Order Logic in Distributive Normal Form.* Ph.D. thesis, University of Cape Town.

NOVAES, C. D. 2020. *The Dialogical Roots of Deduction: Historical, Cognitive, and Philosophical Perspectives on Reasoning.* Cambridge: Cambridge University Press.

PARDO, P. 2022. A modal view on resource-bounded propositional logics. *Studia Logica*, **110**(4), 1035–1080.

PARIKH, R. 2008. Sentences, belief and logical omniscience, or what does deduction tell us? *The Review of Symbolic Logic*, **1**(4), 459–476.

POPPER, K.R. 1959. *The Logic of Scientific Discovery [1934]*. London: Hutchinson.

PRAKKEN, H. 2005. Coherence and flexibility in dialogue games for argumentation. *Journal of Logic and Computation*, **15**, 1009–1040.

PRAWITZ, D. 1965. *Natural Deduction: A Proof-Theoretical Study*. Uppsala: Almqvist & Wilksell.

PRIMIERO, G. 2009. An epistemic logic for becoming informed. *Synthese*, **167**(2), 363–389.

PROOPS, I. 2005. Kant's conception of analytic judgment. *Philosophy and Phenomenological Research*, **70**(3), 588–612.

PROUST, J. 1989. *Questions of Form: Logic and the Analytic Proposition from Kant to Carnap*. Minneapolis: University of Minnesota Press.

QUINE, W.V.O. 1951. Two dogmas of empiricism. *The Philosophical Review*, **60**, 20–43. Reprinted in Quine (1961).

QUINE, W.V.O. 1961. Two dogmas of empiricism [1951]. *Pages 20–46 of: From a Logical Point of View*. Cambridge, MA: Harvard University Press.

QUINE, W.V.O. 1974. *The Roots of Reference*. Open Court.

QUINE, W.V.O. 1991. Two dogmas in retrospect. *Canadian Journal of Philosophy*, **21**(3), 265–274.

RANTALA, V., & TSELISHCHEV, V. 1987. Surface information and analyticity. *Pages 77–90 of:* BOGDAN, R.J. (ed), *Jaakko Hintikka: A Profile*. Dordrecht: D. Reidel Publishing Co.

REITER, R. 1980. A logic for default reasoning. *Artificial Intelligence*, **13**, 81–132.

RORTY, R. (ed). 1967. *The Linguistic Turn: Essays in Philosophical Method*. Chicago: University of Chicago Press.

RUMFIT, I. 2000. "Yes" and "No". *Mind*, **109**, 781–823.

SANDQVIST, T. 2009. Classical logic without bivalence. *Analysis*, **69**, 211–218.

SAVAGE, L. 1967. Difficulties in the theory of personal probability. *Philosophy of Science*, **44**, 305–310.

SEQUOIAH-GRAYSON, S. 2008. The scandal of deduction. Hintikka on the information yield of deductive inferences. *The Journal of Philosophical Logic*, **37**(1), 67–94.

SEXTUS. 1933. *Outlines of Pyrrhonism*. Cambridge (Mass.): Harvard University Press. Translated by R.G. Bury.

SHANNON, C.E., & WEAVER, W. 1949. *The Mathematical Theory of Communication*. Urbana: University of Illinois Press. Foreword by R.E. Blahut and B. Hajek.

SHEERAN, M., & STÅLMARCK, G. 2000. A tutorial on Stålmarck's proof procedure for propositional logic. *Formal Methods in System Design*, **16**, 23–58.

SILLARI, G. 2008a. Models of awareness. *Pages 209–240 of:* BONANNO, G., VAN DER HOEK, W., & WOOLDRIDGE, M. (eds), *Logic and the Foundations of Games and Decisions*. University of Amsterdam.

SILLARI, G. 2008b. Quantified logic of awareness and impossible possible worlds. *Review of Symbolic Logic*, **1**(4), 1–16.

SIM, K.M. 1997. Epistemic logic and logical omniscience: A survey. *International Journal of Intelligent Systems*, **12**(1), 57–81.

SMILEY, T. 1996. Rejection. *Analysis*, **56**, 1–9.

SMULLYAN, R. 1968. *First-Order Logic*. Berlin: Springer.

SOLARES-ROJAS, A. 2022. *Tractable Depth-Bounded Approximations to Some Propositional Logics*. Ph.D. thesis, University of Milan.

STATMAN, R. 1979. Intuitionistic propositional logic is polynomial-space complete. *Theoretical Computer Science*, **9**, 67–72.

STOCKMEYER, L. 1987. Classifying the computational complexity of problems. *Journal of Symbolic Logic*, **52**(1), 1–43.

STÅLMARCK, G. 1992. *A system for determining propositional logic theorems by applying values and rules to triplets that are generated from a formula*. Swedish Patent No. 467 076 (approved 1992), U.S. Patent No. 5 276 907 (1994), European Patent No. 0403 454 (1995).

STRASSER, C., & ANTONELLI, G.A. 2019. Non-monotonic logic. *In:* ZALTA, E.N. (ed), *The Stanford Encyclopedia of Philosophy*, Summer 2019 edn. Metaphysics Research Lab, Stanford University.

TENNANT, N. 1984. Perfect validity, entailment and paraconsistency. *Studia Logica*, **43**, 179–198.

TENNANT, N. 1987. Natural deduction and sequent calculus for intuitionistic relevant logic. *Journal of Symbolic Logic*, **52**(3), 665–680.

TENNANT, N. 1990. *Natural Logic*. Edinburgh: Edinburgh University Press.

THANG, P.M., & LUONG, H.T. 2014. Translating preferred subtheories into structured argumentation. *Journal of Logic and Computation*, **74**(4), 831–849.

URQUHART, A. 1984. The undecidability of entailment and relevant implication. *Journal of Symbolic Logic*, **49**(4), 1059–1073.

URQUHART, A. 1990. The complexity of decision procedures in relevance logic. *Pages 61–76 of:* E A. GUPTA, J.M. DUNN (ed), *Truth or Consequences*. Kluwer Academic Publishers.

URQUHART, A. 1995. The complexity of propositional proofs. *Bulletin of Symbolic Logic*, **1**(4), 425–467.

VAN BERKEL, K., D'AGOSTINO, M., & MODGIL, S. 2022. A Dialectical Formalisation of Preferred Subtheories Reasoning Under Resource Bounds. *Forthcoming*.

VLASTOS, G. 1982. The Socratic elenchus. *The Journal of Philosophy*, **79**(11), 711–714.

WITTGENSTEIN, L. 2001. *Tractatus Logico-Philosophicus [1921]*. London–New York: Routledge.

YATO, T., & SETA, T. 2003. Complexity and completeness of finding another solution and its application to puzzles. *IEICE TRANSACTIONS on Fundamentals of Electronics, Communications and Computer Sciences*, **86**(5), 1052–1060.

YOUNG, A.P., MODGIL, S., & RODRIGUES, O. 2016. Prioritised default logic as rational argumentation. *Pages 626–634 of: Proceedings of the Fifteenth International Conference on Autonomous Agents and Multiagent Systems (AAMAS '16)*. Red Hook, NY: Curran Associates Inc.

Index

0-depth consequence, 34, 41, 64
0-depth deducible, 41
0-depth inconsistency, 34, 63
0-depth logic, 34, 61
    tractability, 54
$\Vdash_k^\Lambda$, 73
$\vDash_k^g$, 74
p-simulation, 80

at, 70, 71
a posteriori, 137, 143
a priori, 137, 142, 143, 165, 174
actual assumptions, 76
actual information, xxiv, 9, 10, 12, 15, 16, 23–26
admissible
    extension, 113, 117, 122–124, 130, 132, 133
    semantics, 113
    sets of arguments, 112
analytic
    argument, *see* analytic, inference
    inference, 14–16, 31, 146, 163–165, 167, 168, 171, 172, 174–176, 179
    in the syntactic sense, 52
    judgment, *see* analytic, statement
    logic is, 154, 160, 161, 170, 171
    proof, *see* analytic, inference
    proposition, *see* analytic, statement
    sentence, *see* analytic, statement
    statement, 137–145, 147, 157, 161, 163, 171, 175
    truth, *see* analytic, statement
analytic-synthetic distinction, 137–139, 141, 143, 144, 165, 166, 170, 171, 173, 174, 176
    Carnap's, 144
    Frege's, 140–141
    Hintikka's, 165–171
    Kant's, 138–139
    logical empiricism's, 143
    Quine's, 143–146
analyticity, 137, 138, 140–147, 162, 164, 165, 171–173,

175
approximation, xi, xii, 8, 69,
        71–73, 176, 186, 188
    depth-bounded, 13, 68–70,
        73, 75, 83, 89, 101, 103
      strong, 73, 87, 88, 93, 94,
        96
      tractable, 187
      weak, 65, 73, 74, 93
    of classical meaning, 29
    problem, xii, xiv, 12, 181
    relation, 28, 46
    system, 7–9, 12–14, 24, 65,
        68, 70, 71, 73, 78, 88,
        96, 97
    tractable, 73
approximation system
    properties, 8
argument game, 118, 122
    proof theories, 117, 118, 136
argumentation, 110
    Dung's theory of, 111
argumentation based inference,
        115, 117, 118
argumentation framework, 112,
        114
    k-depth, 127, 130

Bar-Hillel and Carnap paradox,
        153, 178
Belnap's 4-valued logic, 5

C-intelim
    hyper-sequence, 98
    tableaux, 98
C-intelim tableaux, 77
C-intelim hyper-proof, 90

C-intelim hyper-sequence, 88, 91
    k-depth, 89
    k-depth, 96
    non-redundant, 92
    strongly non-redundant, 92
C-intelim tableau, 77
C-intelim tableaux, 75, 78, 90, 97
    atomically normal, 82
    g-normal, 82
    non-redundant, 80
    normal, 80
    proper, 102
    quasi-normal, 80, 83
    strongly non-redundant, 81
C-intelim$^+$, 90, 91
categorical judgments, 138–141,
        166
categorical propositions, *see*
        categorical judgments
classical logic based arguments,
        111
closure rule, 77
Cohen-Nagel paradox, 147, 172
committments, 127–130
complete
    extension, 113, 122–124,
        130
    semantics, 113
conception of logic
    Ayer's, 157–158
    Carnap's, 144
    Frege's, 140–142, 155
    Hahn's, 155–156
    Hempel's, 157
    Kant's, 139
    logical empiricism's, 143

Quine's, 144–145
Wittgenstein's, 160
conflict free sets of arguments, 112, 113, 122, 123, 129, 130
coNP-complete, 4, 5
consensus principle, xxi, xxiv, 9–11
Consistency, 122–124, 132, 133
consistency checking, 110, 124, 125, 127, 130
constituent, 167–168, 170, 171
    non-trivially inconsistent, 167, 168
    subordinate, 167
    trivially inconsistent, 167, 168
contaminated proof, 101
contamination problem, 99
cut, 34, 53, 78
    analytic, 53
    bounded, 88

defeasible inference, 108–111, 115, 119, 135
defeasible reasoning, *see* defeasible inference
detour, 50–52, 81, 93
dialectical
    acceptability, 125, 127
    argument, 132, 133
    classical logic
        argumentation, 119, 131–133
    defeat, 128
    extension, 130, 132
    reasoning, 119, 125, 127–130, 134
dialogue, 110, 111, 117, 118, 122, 127
    graph, 118
distributive normal forms
    Hintikka's theory of, 166–170

elementary argument, 112, 129
epistemic variant, 128, 130, 132, 133, 135
ex-falso, 48, 100
explosively contaminating, 133
explosivity, 34, 48–50
expressive completeness, 63–64

F, 72
Frege's concept formation, 162
Frege's function-argument analysis, 141
Frege's logicism, 141, 161

g-analytic
    application of RB, 77, 82, 83, 98
grounded
    extension, 113, 114, 116, 117, 124, 125
    semantics, 113, 117

Hintikka's depth
    of a first-order constituent, 167–168
    of a first-order formula, 166–168
Hintikka's depth information, 169, 174

Hintikka's surface information, 169–170, 174
hypothetical reasoning, 67

intelim*, 47
idle formula, 51, 81
implicit information, 3, 4, 12, 14, 16, 29, 31, 46, 58, 65
implicitly contained, *see* information containment
improper proof, 77, 101, 102
information containment, 3, 4, 9, 10, 14, 29, 31, 46, 67
information state, 10, 22–25, 31
    0-depth, *see* information state, shallow
    k-depth, 87
    shallow, 31–34, 43, 46, 53, 54, 65–69, 73–75, 87, 89
informational content, 142, 153
informational indeterminacy, 17, 18
informational meaning of the logical operators, *see* informational semantics
informational semantics, xiv, 13, 14, 16–18, 22, 24, 25, 63, 121, 174, 176, 181, 185
    via negative constraints, 25, 26, 31, 54, 55, 68
informational truth and falsity, 16–18, 20, 22, 23, 25
informational values, 26

informational view of logic, xiv, 3, 13
informationally trivial, xii, xiii, 3–5, 7, 12, 14–16, 24, 147, 172, 176, 178, 179
input formulae, *see* input set
input set, 70, 72, 78, 83, 97
intelim
    deducibility, 41
    deducible, 37, 41, 52
    inconsistent, 37, 41, 48
    proof, 37, 41, 48–50, 52
        subformula property of, 47, 48
    proofs, 53
    refutable, 37, 41
    refutation, 37, 48, 52, 53
        subformula property of, 47, 48
    rules, 36, 37, 41, 43, 46, 50, 53, 65
    sequence, 35, 37, 41, 43, 44, 48–50, 52, 65, 76–78
        atomically closed, 41
        closed, 37
        non-redundant, 51
        strongly non-redundant, 52
intelim sequence, 76
intelimi
    sequence
        closed, 77
inter-agent joint deliberation, 118
intuitionistic logic, 5, 11, 12, 22, 34, 120
    Beth's semantics for, 23

Kripke semantics for, 23
inversion principle, 50

Kant's containment criterion, 138, 139, 141
KE, 80
KI, 80
Kleene's 3-valued logic, 18, 19, 22, 34, 187

$\mathcal{L}$-domain, 53
$\mathcal{L}$-module, 26, 54, 58
linguistic turn, 142
locutions, 110, 111, 118, 119
logical empiricism, 137, 142, 143, 147, 155, 173
logical laws, *see* logical truths
logical omniscience, xi, xiii, 6, 7, 110, 120, 122, 123, 134, 149–151, 186
logical system, xiii, 7
logical truths, 139, 140, 144, 145, 148, 149, 153
logically omniscient, *see* logical omniscience
logically perfect language, 158–160

many-valued logic, 34
matrix, *see* valuation system
maxiconsistent
   approaches, 110, 111, 115, 134
   non-monotonic
      consequence, 111
   non-monotonic logics, 115
   paradigm, 122, 135

minimal rationality, 121, 122, 134
monotonicity of classical logic, 108

natural deduction, xiv, 10–12, 51, 77, 97, 101, 102, 120, 135, 182–185
non-categorical judgments, 138, 166
non-contamination, 98, 122, 124, 125, 133, 135
non-deterministic semantics, *see* valuation, non-deterministic
non-monotonic logics, 107–111, 115, 117, 118, 130, 133–136
non-monotonic reasoning, xiv, xv, 5, 107, 108, 111, 115, 117–119, 121, 122, 125, 134
   distributed, 111
   resource-bounded, 119
NP-complete, xviii, 5

omniscience, 17, 32, 66, 155, 156

P $\neq$ NP, 12, 160
p-simulation, 78, 80
paradox of analysis, 147
partial valuation, 65, 66, 177
Peirce's arrow, 27
philosophy of logic, xv, 137–179
Pickwickian proof, 49, 50, 52
polynomially bounded, 71–73, 78

preference relation, 108, 112, 113, 115, 123
   elitist, 112, 115, 117
preferred
   extension, 113–117, 130
   semantics, 113, 117
Preferred Subtheories, 108, 115, 131, 133
Premise Consistency, 123, 124, 132, 133
Principle of Bivalence, 15, 20, 182
   informational version of, 13, 17, 32, 89
Principle of Non-Contradiction, 15, 139
   informational version of, 13, 17, 89
proof boxes, 90, 91
PSPACE-complete, 5, 12

R-contaminated argument, 125, 135
R-contaminated proof, 101, 103, 133
rationality postulates, 122, 135
RB rule, 76–78, 82, 91, 97
reasoning by cases, 66
reductio ad absurdum, 68
refinement, 27, 66
   strict, 28
resource-bounded
   agent, 107, 134
      non-monotonic reasoning, 119
      rationality, 119

sub, 70–72, 78
sub_k, 72
scandal of deduction, 147–149, 155, 161, 171
SCP, xviii, xx, xxi, xxiv, 9, 12, 29, 31–33, 35, 44, 46, 47, 54–56, 58, 65–69, 74
search space, 69–73, 97
   canonical, 72, 73, 84, 86, 87, 93, 94, 96, 98
   function, 70–72, 78
   canonical, 72
semantic information, 151, 153, 154, 169, 176, 178
Sheffer's stroke, 27
signed formula, 33
Single Candidate Principle, *see* SCP
Smullyan's tableaux, 77, 80
Socratic move, 127
speech act, 119
stable
   extension, 113–116, 130
   semantics, 113
strictly analytic
   inference, 172
strictly tautological
   inference, 172
subformula
   weak, 53
subformula graph, 54, 85
subformula property, 47, 49, 52
   generalized, 77, 84, 93, 94
   weak, 53, 84
subset minimality, 124

sudoku, xv, xvii, xviii, xxi, xxiv, 9, 29, 120
suppositions, 128, 131, 132
syntactically disjoint
    formulae, 99
    sets, 99
synthetic
    a priori, 143, 165, 174
    argument, 68, *see* synthetic, inference
    inference, 163, 165–168, 171, 174, 175
    judgments, *see* synthetic, statement
    proposition, *see* synthetic, statement
    statement, 137–143, 145, 165, 166, 175

Tarskian consequence relation, *see* Tarskian logic
Tarskian logic, 34, 65, 88, 97
    properties, 34
tautological, 148, 149, 151, 161, 170, 171
    in the informational sense, 146, 147, 172
    logic is, 146, 154, 160, 165, 168
tractability, 86, 97
truth tables, 15, 17, 18, 65
truth-table, 80
truth-tables, 78, 80

unconditional argument, 130, 132
undercut attack, 131

valuation, 28, 58, 65, 66
    3ND, 62
    admissible, 27, 28, 32, 46
    Boolean, 19, 20, 32, 61
    inadmissible, 28
    module, 26, 27, 29, 31, 35, 44, 46, 55, 56, 65–67
        admissible, 27
        closed under SCP, 67
        inadmissible, 27, 46
        stable, 29, 31, 32, 46, 54–56, 58, 59, 66–68
        unstable, 29, 32, 33, 55, 56, 58, 59, 67, 68, 74
    non-deterministic, 59, 61–63
    partial, 17, 19, 20, 25–28, 35, 46, 53, 63
    system, 59–61, 63
        lack of a finite, 60
valuation module
    closed under SCP, 66
variable sharing, 103
veridicality thesis, 177
Vienna Circle, 142, 143, 148, 171
virtual space
    function, 75
virtual assumptions, 11, 76, 77, 81, 84, 91, 184
virtual information, xvii, xxi, xxiv, 9–11, 13, 24, 35, 65–69, 76, 90, 120, 121, 134, 174–176, 182, 186, 187
virtual space, 69–71, 77
    function, 70–73, 90

www.ingramcontent.com/pod-product-compliance
Lightning Source LLC
Chambersburg PA
CBHW071708160426
43195CB00012B/1616